NJU SA 2015-2016

THE YEAR BOOK OF ARCHITECTURE PROGRAM SCHOOL OF ARCHITECTURE AND URBAN PLANNING

南京大学建筑与城市规划学院建筑系 教学年鉴

王 丹 丹 编 EDITOR: WANG DANDAN

东南大学出版社·南京 SOUTHEAST UNIVERSITY PRESS, NANJING

建筑设计及其理论
Architectural Design and Theory

张 雷 教 授	Professor ZHANG Lei
冯金龙 教 授	Professor FENG Jinlong
吉国华 教 授	Professor JI Guohua
周 凌 教 授	Professor ZHOU Ling
傅 筱 教 授	Professor FU Xiao
钟华颖 讲 师	Lecturer ZHONG Huaying

城市设计及其理论
Urban Design and Theory

丁沃沃 教 授	Professor DING Wowo
鲁安东 教 授	Professor LU Andong
华晓宁 副教授	Associate Professor HUA Xiaoning
胡友培 副教授	Associate Professor HU Youpei
窦平平 副教授	Associate Professor DOU Pingping
刘 铨 讲 师	Lecturer LIU Quan
尹 航 讲 师	Lecturer YIN Hang
唐 莲 讲 师	Lecturer TANG Lian

建筑历史与理论及历史建筑保护
Architectural History and Theory,
Protection of Historic Building

赵 辰 教 授	Professor ZHAO Chen
王骏阳 教 授	Professor WANG Junyang
胡 恒 教 授	Professor HU Heng
冷 天 讲 师	Lecturer LENG Tian
王丹丹 讲 师	Lecturer WANG Dandan
刘 妍 讲 师	Lecturer LIU Yan

建筑技术科学
Building Technology Science

鲍家声 教 授	Professor BAO Jiasheng
秦孟昊 教 授	Professor QIN Menghao
吴 蔚 副教授	Associate Professor WU Wei
郜 志 副教授	Associate Professor GAO Zhi
童滋雨 副教授	Associate Professor TONG Ziyu

南京大学建筑与城市规划学院建筑系
Department of Architecture
School of Architecture and Urban Planning
Nanjing University
arch@nju.edu.cn http://arch.nju.edu.cn

教学纲要

EDUCATIONAL PROGRAM

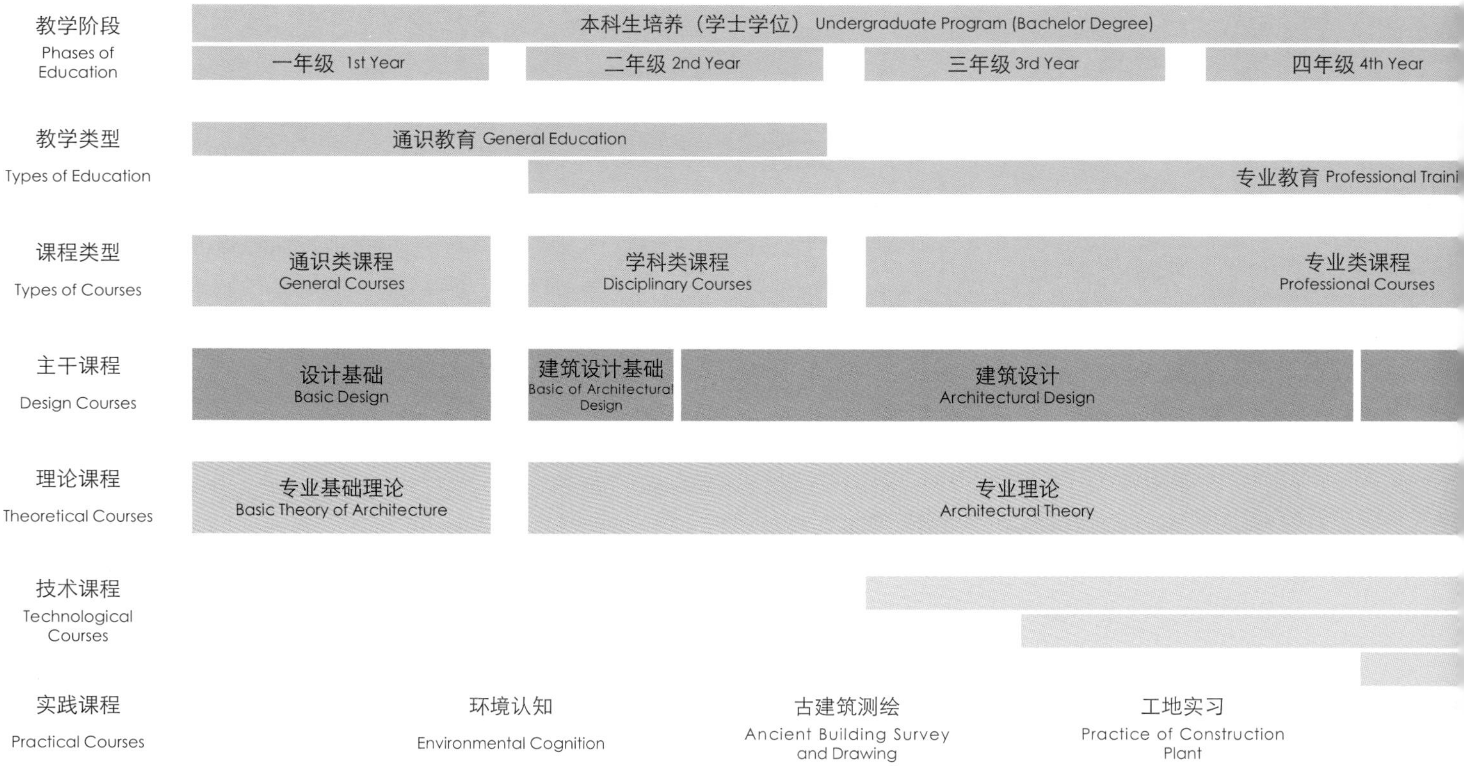

教学阶段 Phases of Education	本科生培养（学士学位） Undergraduate Program (Bachelor Degree)			
	一年级 1st Year	二年级 2nd Year	三年级 3rd Year	四年级 4th Year
教学类型 Types of Education	通识教育 General Education			
		专业教育 Professional Traini		
课程类型 Types of Courses	通识类课程 General Courses	学科类课程 Disciplinary Courses	专业类课程 Professional Courses	
主干课程 Design Courses	设计基础 Basic Design	建筑设计基础 Basic of Architectural Design	建筑设计 Architectural Design	
理论课程 Theoretical Courses	专业基础理论 Basic Theory of Architecture	专业理论 Architectural Theory		
技术课程 Technological Courses				
实践课程 Practical Courses	环境认知 Environmental Cognition	古建筑测绘 Ancient Building Survey and Drawing	工地实习 Practice of Construction Plant	

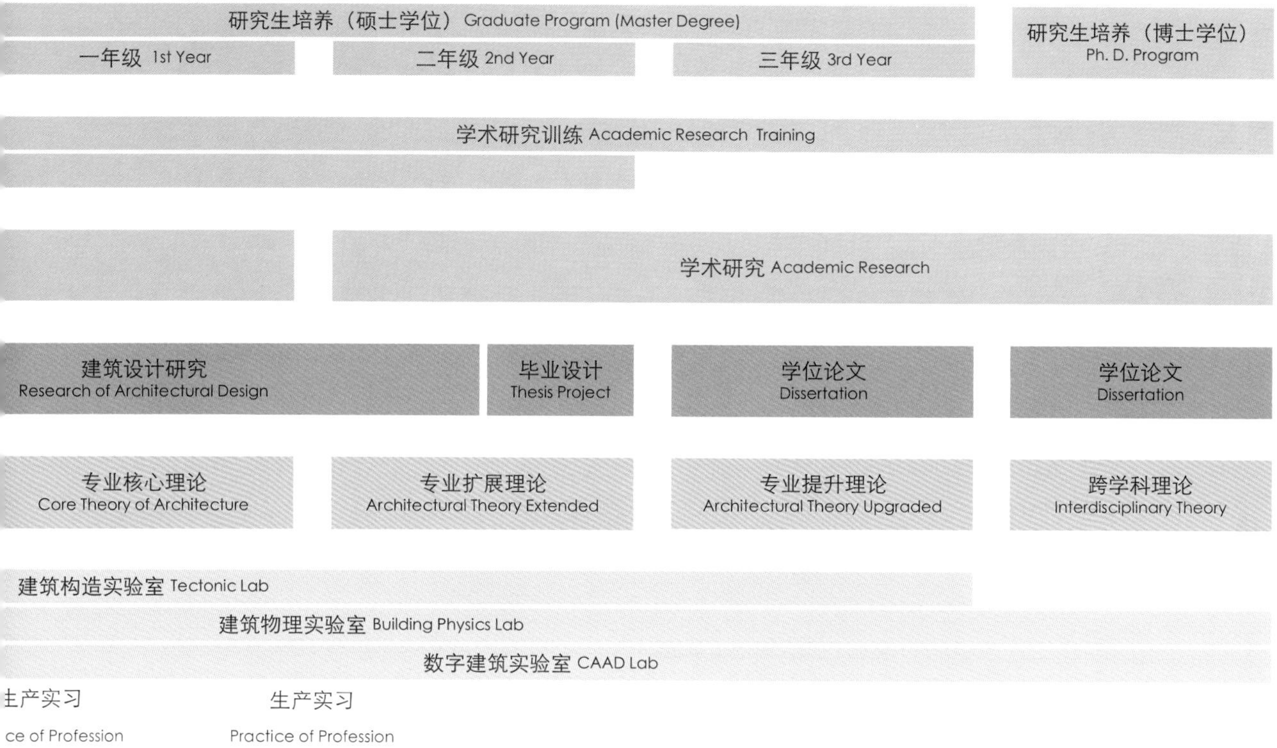

研究生培养（硕士学位）Graduate Program (Master Degree)

一年级 1st Year

二年级 2nd Year

三年级 3rd Year

研究生培养（博士学位）
Ph. D. Program

学术研究训练 Academic Research Training

学术研究 Academic Research

建筑设计研究
Research of Architectural Design

毕业设计
Thesis Project

学位论文
Dissertation

学位论文
Dissertation

专业核心理论
Core Theory of Architecture

专业扩展理论
Architectural Theory Extended

专业提升理论
Architectural Theory Upgraded

跨学科理论
Interdisciplinary Theory

建筑构造实验室 Tectonic Lab

建筑物理实验室 Building Physics Lab

数字建筑实验室 CAAD Lab

生产实习
ce of Profession

生产实习
Practice of Profession

课程安排
CURRICULUM OUTLINE

	本科一年级 Undergraduate Program 1st Year	本科二年级 Undergraduate Program 2nd Year	本科三年级 Undergraduate Program 3rd Year
设计课程 Design Courses	设计基础 Basic Design	建筑设计基础 Basic Design of Architecture 建筑设计（一） Architectural Design 1 建筑设计（二） Architectural Design 2	建筑设计（三） Architectural Design 3 建筑设计（四） Architectural Design 4 建筑设计（五） Architectural Design 5 建筑设计（六） Architectural Design 6
专业理论 Architectural Theory	逻辑学 Logic	建筑导论 Introductory Guide to Architecture	建筑设计基础原理 Basic Theory of Architectural Design 居住建筑设计与居住区规划原理 Theory of Housing Design and Residential Planning 城市规划原理 Theory of Urban Planning
建筑技术 Architectural Technology	理论、材料与结构力学 Theoretical, Material & Structural Statics Visual BASIC程序设计 Visual BASIC Programming	CAAD理论与实践 Theory and Practice of CAAD	建筑技术（一） 结构与构造 Architectural Technology 1: Structure & Construction 建筑技术（二） 建筑物理 Architectural Technology 2: Building Physics 建筑技术（三） 建筑设备 Architectural Technology 3: Building Equipment
历史理论 History Theory	古代汉语 Ancient Chinese	外国建筑史（古代） History of World Architecture (Ancient) 中国建筑史（古代） History of Chinese Architecture (Ancient)	外国建筑史（当代） History of World Architecture (Modern) 中国建筑史（近现代） History of Chinese Architecture (Modern)
实践课程 Practical Courses		古建筑测绘 Ancient Building Survey and Drawing	工地实习 Practice of Construction Plant
通识类课程 General Courses	数学 Mathematics 语文 Chinese 名师导学 Guide to Study by Famed Professors 计算机基础 Basic Computer Science	社会学概论 Introduction of Sociology 社会调查方法 Methods for Social Investigation	
选修课程 Elective Courses		城市道路与交通规划 Planning of Urban Road and Traffic 环境科学概论 Introduction of Environmental Science 人文科学研究方法 Research Method of the Social Science 美学原理 Theory of Aesthetics 管理学 Management 概率论与数理统计 Probability Theory and Mathematical Statistics 国学名著导读 Guide to Masterpieces of Chinese Ancient Civilization	人文地理学 Human Geography 中国城市发展建设史 History of Chinese Urban Development 欧洲近现代文明史 Modern History of European Civilization 中国哲学史 History of Chinese Philosophy 宏观经济学 Macro Economics 管理信息系统 Management Operating System 城市社会学 Urban Sociology

本科四年级 Undergraduate Program 4th Year	研究生一年级 Graduate Program 1st Year	研究生二、三年级 Graduate Program 2nd & 3rd Year
建筑设计（七） Architectural Design 7 建筑设计（八） Architectural Design 8 本科毕业设计 Graduation Project	建筑设计研究（一） Design Studio 1 建筑设计研究（二） Design Studio 2 数字建筑设计 Digital Architecture Design 联合教学设计工作坊 International Design Workshop	专业硕士毕业设计 Thesis Project
城市设计理论 Theory of Urban Design	城市形态研究 Study on Urban Morphology 现代建筑设计基础理论 Preliminaries in Modern Architectural Design 现代建筑设计方法论 Methodology of Modern Architectural Design 景观都市主义理论与方法 Theory and Methodology of Landscape Urbanism	
建筑师业务基础知识 Introduction of Architects' Profession 建设工程项目管理 Management of Construction Project	材料与建造 Materials and Construction 中国建构（木构）文化研究 Studies in Chinese Wooden Tectonic Culture 计算机辅助技术 Technology of CAAD GIS基础与运用 Concepts and Application of GIS	
	建筑理论研究 Study of Architectural Theory	
生产实习（一） Practice of Profession 1	生产实习（二） Practice of Profession 2	建筑设计与实践 Architectural Design and Practice
景观规划设计及其理论 Theory of Landscape Planning and Design 东西方园林 Eastern and Western Gardens 地理信息系统概论 Introduction of GIS 欧洲哲学史 History of European Philosophy 微观经济学 Micro Economics 政治学原理 Theory of Political Science 社会学定量研究方法 Quantitative Research Methods in Sociology	建筑史研究 Studies in Architectural History 建筑节能与可持续发展 Energy Conservation & Sustainable Architecture 建筑体系整合 Advanced Building System Integration 规划理论与实践 Theory and Practice of Urban Planning 景观规划进展 Development of Landscape Planning	

1—157 年度改进课程 WHAT'S NEW

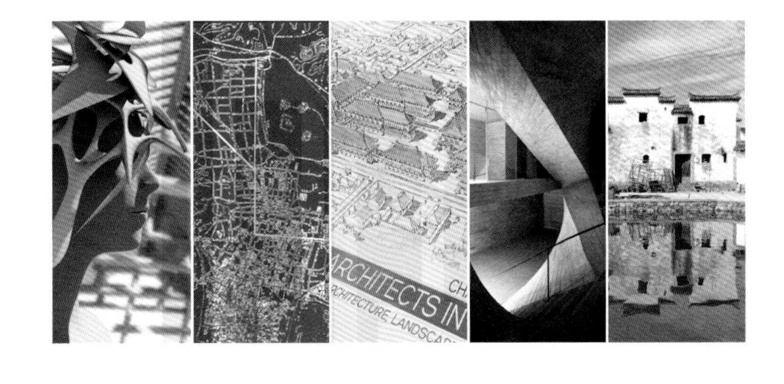

ARCHITECTS IN
CH

年度改进课程

WHAT'S NEW

设计基础(二)

BASIC DESIGN 2

丁沃沃 鲁安东 唐莲

南京大学建筑学本科自2007年设立开始，本科一年级一直以通识教育为主。通识教育夯实学生的知识基础，包括文科、理科与美学三方面的课程。美学课程与南京艺术学院合作开展，第一学期进行视觉训练，第二学期进行空间训练。建筑空间以人为本，空间训练强调以“身体”为核心进行课程的设置，共包括三个部分的练习。练习一“动作-空间分析”通过分析被空间限定的身体动作，训练学生认知身体、尺度与环境的关系；练习三“互承的艺术”通过真实搭建身体能够进入或通过的空间结构，建立学生对建筑结构的初步认识；练习二在前两年“折纸空间”的基础上，将纯粹对纸的操作转化为与身体关系更为紧密的“空间包裹”，训练学生形成建筑学形式操作的基本思维与方法，更系统地衔接练习一与练习三。

1. 课程设置

“空间包裹——折纸的艺术”的教学历时五周，要求用折纸对身体的一个部位进行包裹，完成一件衣服的设计与制作。课程可以理解为基于身体（场地）的形式操作，教学的主要内容是形式设计的逻辑与方法，其中折纸作为实现形式的技术与媒介。为此，在整个教学过程中设置了三个阶段的练习，并开展相应的讲座来指导与配合练习。这三个阶段分别为，折纸单元基础练习（一周）、折纸单元的变形与组合研究（一周）以及折纸包裹空间设计（三周）。

1.1折纸单元基础练习

阶段一折纸单元基础练习训练学生对材料、形式单元的认知。学生需学习折纸的基本知识，运用单元拼插或者整纸折叠的方式，制作一个直径不小于15cm的空心球。这个练习有助于学生快速掌握折纸技术，了解形式单元与基本形——球之间的构成关系。

材料认知是建筑学一项重要的训练内容，折纸练习中，通过对白纸的折叠以及白纸形成构件单元与单元之间的拼接，学生切身体会了材料与工艺、构件、结构等的关系。白纸的厚薄、质感等特性直接关系到球的制作是否能够成功。另外，从平面的白纸到三维的球的过程，初步训练了学生对形式单元与最终形式关系的理解。球是最简单的空间包裹体，各个方向的弧度完全一致，只需要有规律地重复折纸单元就能完成球的制作。球制作完成后，学生被要求对使用纸张的种类、大小、数量、构成球的构件单元、拼接方式、单元数量等进行统计，将折纸单元尺寸及数量与球的弧度建立链接，最后与其他同学的成果进行比较，分析形式单元的塑形效率。

1.2折纸单元的变形与组合研究

阶段二折纸单元的组合与变形研究训练学生掌握形式变形规律，培养理性的思维方法。该练习要求学生以折纸球的单元为基础选择一种单元进行深入，单元拼插通过单元大小组合、整纸折叠通过折痕线的变化，来研究折纸塑形机制，最终能够做到娴熟地控形。这个练习在整个教学过程中非常关键，鼓励学生学习、探索与研究，在掌握老师传授的基本原理和知识基础上，学生需要自己选择合理的可发展的折纸单元，对单元进行改进（以做到更稳固的连接或完成更丰富的变化），探索选定单元的组合与变化的所有可能性及适用性，并能够图解清晰的数形关系。学生对塑形机制掌握得越充分，在下一个练习中就能够越娴熟地进行成品设计。

对建筑设计过程的关注，以及对形式生成原因的研究在建筑学训练中越来越被重视。本阶段练习中学生需探索与分析单元操作与形式变化的关系。折纸从平面到三维形式的规律可以通过简单的几何知识进行归纳。学生在练习过程中，除了通过模型来呈现变形与组合的可能之外，还需手绘图解单元几何尺寸、关系，对变形和组合的原理进行图示解析，了解折纸操作中单元的变形、不同组合、峰折、谷折等对于形式控制的意义。

1.3折纸包裹空间设计

阶段三折纸包裹空间设计训练学生对形式规律的运用及设计能力，要求学生运用掌握的折纸塑形原理，包裹身体的一个部位，最终成品需满足身体尺度的三个层次，并能够改变人的形体。这个过程历时三周，前两周以设计与制作衣服为主，学生需要根据选择的部位以及衣服的设计意向设定概念，设计除了在技术上遵循之前的研究成果之外，也需符合概念的设定，由概念来引导设计造型的走向；后一周学生需要对成品进行拍摄，对制作原理进行手绘图表述，最终完整呈现到一张图版上。

建筑形式不仅仅是一种造型，形式与场地关系密切，需具备合理性，课程要求衣服的设计需顺应身体尺度的需要。身体作为折纸衣服的“场地”，具有三个层次的尺度。首先，基本尺度，人的身体可以理解为多个球面体或多面体的组合，不同部位具有不同的尺寸，且不同部位弧度存在差异。其次，穿戴尺度，人的身体是可活动的，穿戴与活动的最大尺寸、最小尺寸决定了衣服尺寸的可变区间。第三，扩展尺度，在满足前两个层次的基础上，身体的形体可被衣服重塑与改变，还具备可扩展尺度。最终，衣服成品的形式控制是否完成了身体尺度的三个层次、是否遵从折纸单元的塑形

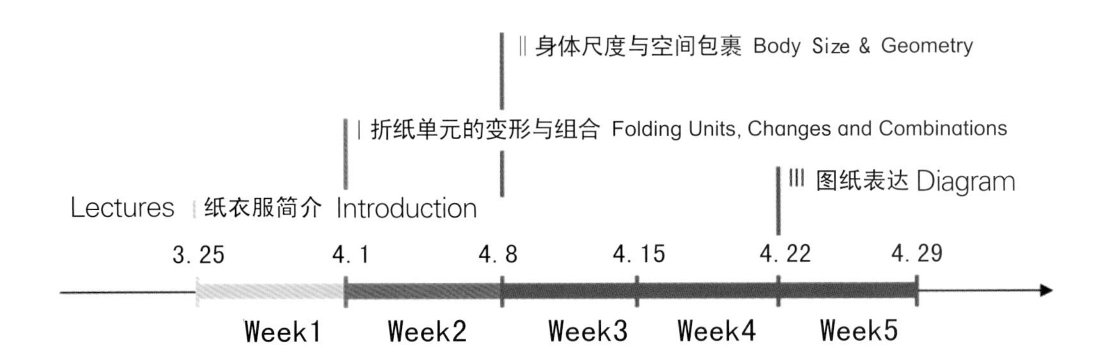

机制以及是否遵从概念的设定，都是评判作品是否优秀的考量因素。

2. 教学成果与讨论

“空间包裹——折纸的艺术”课程取得了较好的教学效果，学生兴趣浓厚，最终作品及图纸完成度较高。由于理性思考与控制的存在，“可以复制”成为最终作品重要的特点，也因此很好地贴合了课程设置的训练目标。对于一学期的空间基础教学来说，“空间包裹”承接了“动作-空间分析”中学生对身体尺度及空间关系的认知，培养了学生理性的形式操作能力，为接下来的“互承的艺术”的真实搭建打下了扎实的基础。学期末“艺术的理性”设计成果展中，学生穿着设计制作的折纸服装，穿梭于亲自搭建的覆盖结构之中，再次感受到材料之美、结构之美、空间之美。至此，三个课程完成了空间基础的整体训练。

“纸”与“身体”作为艺术学院的经典训练项目，强调对服装设计的训练；在本教案中，当“纸”与“身体”变为建筑学训练的载体时，则强调形式构成与形式逻辑训练。“空间包裹”练习中含有建筑学中的多个基本问题，比如场地问题、材料问题、构造问题、结构问题、设计问题等。将身体看做场地，折纸衣服看做建筑，最终完成的折纸衣服既是独立的作品，更是借以思考建筑学问题的载体。对于建筑学来说，好的建筑不仅最终的形式是美的，而且形式的生成应该是理性的，构成形式的构件单元应该是合理的，构件与构件之间的拼插应该是严谨的，这些都涵盖在设计过程中。加强思维逻辑训练，通过设计过程训练设计思维，将延续到建筑学专业训练的全过程。

Since the major of architecture was established at Nanjing University in 2007, general education has always been the focus for first-year undergraduate students . General education can enhance knowledge base for students, which involves courses in three aspects of liberal arts, science and aesthetics. The course of aesthetics is carried out jointly with Nanjing University of the Arts. The first semester is for visual training, and the second semester is for space training. Architectural space is human oriented, so course setup for space training is carried out by emphasizing “human body” as the core, and it consists of three exercises. Exercise 1 is “action-space analysis”, which aims to train students to understand the relations among human body, dimensions and environment through analysis on actions of human body confined by space; Exercise 3 is “art of mutually-supporting”, which aims to help students establish preliminary understanding on architectural structure through erecting a physical structure of space that can be assessed or gone through by human body; Exercise 2 transforms pure paper operation into “spatial enclosure” that has closer connection with human body on basis of the “folding space” established two years ago, and is intended to train students to shape basic thought and methods for operation of architectural forms, and to link Exercise 1 and Exercise 3 more systematically.

1.Setup of the Course

The teaching process of “spatial enclosure – the art of paper folding” lasts five weeks; it requires students to enclose one part of human body with paper folding to complete the design and manufacture of a piece of clothing. The course can be construed as form operation based on human body (the site), and main teaching content is the logic and methods of form design, in which paper folding is applied as technique and medium to realize the form. Therefore, three exercise phases are arranged in the entire teaching process, and relevant lectures are unfolded to direct and assist exercises. These three phases are basic exercise of folding units (one week), research on transformation and combination of folding units (one week), and design of space enclosed with paper folding (three weeks).

1.1Basic Exercise of Folding Units

The basic exercise of folding units in Phase 1 aims to train students to understand materials and form units. Students shall learn basic knowledge of paper folding, and make a hollow sphere with diameter no less than 15cm through the way of splicing units or one-paper folding. This exercise can help students grasp folding skills rapidly, and understand the formation relationship between form units and basic shapes–sphere.

Understanding materials is an important item of training content of architecture. During the paper folding exercise, students can experience by themselves the relations among materials, techniques, members, and structure through folding of white paper and splicing of different member units. Thickness, texture and other features of white paper have a direct bearing on success of fabrication of the sphere. In addition, the process from two-dimensional paper to a three-dimensional sphere trains students to understand the relationship between form units and final form. Sphere is the simplest spatial enclosure, with completely identical arc in every direction, and can be completed only by repeating the folding unit regularly. After the sphere is completed, students need to count types, sizes, and quantity of paper used, as well as member units shaping the sphere, splicing mode, and quantity of units, establish links between sizes and quantity of folding units and arc of the sphere and finally, compare with results of other students, and analyze molding efficiency of form units.

1.2Research on Transformation and Combination of Folding Units

The research on transformation and combination of folding units in Phase 2 aims to train students to grasp transformation rules of forms, and to cultivate rational thinking methods. This exercise requires students to select one unit for deeper exercise on basis of the unit of paper folded sphere, study molding mechanism of paper folding through combination of different sizes of units to be spliced or through variation of folding lines for one-paper folding, and realize skilled forming control in the end. This exercise is very important for entire teaching process. Students are encouraged to learn, explore and research, and on basis of basic principles and knowledge basis taught by the teacher, students are required to choose reasonable and developable folding units by themselves, improve such units (so as to realize more stable connection or complete more variations), explore all possibility and applicability of combination and transformation of selected units, and illustrate clear relationship between quantity and forms with drawings. The more sufficiently students master the molding mechanism, the more skillfully they can complete design of finished product during next exercise.

Attention on the process of architectural design, and research on reasons of form shaping are increasingly emphasized during training of architecture. In this phase of exercise, students are required to explore and analyze relations between unit operation and form transformation. The rules of paper folding from two-dimensional to three-dimensional forms can be concluded with geometric knowledge in a simple way. In the process of exercise, in addition to possibility of presenting transformation and combination with models, students are also required to draw geometric dimensions and relations of graphic units manually, illustrate principles of transformation and combination with diagrams, and understand transformation of units during paper folding, as well as the meaning of different combinations, peak folding, and valley folding to form control.

1.3Design of Space Enclosed with Paper Folding

The design of space enclosed with paper folding in Phase 3 aims to train students with the application and design ability for rules of forms. It requires students to apply acquired molding principles of paper folding to enclose one part of human body, and requires that the finished product shall meet three levels of human body dimensions, and can change the shape of human body. This process lasts for three weeks. The earlier two weeks are mainly for clothing design and manufacture, and students are required to set up concepts according to selected body part and intention of clothing design, in addition to following previous research results technically, the design shall also meet setup of concepts, and guide direction of design and molding with concepts; the next one week is for photography of finished products, and students are required to express fabrication principles with hand-drawn drawings, and finally present them completely on one piece of layout.

Architectural form is not just a type of molding, form and site are closely connected, and should be rational, and the course requires that clothing design must meet the demand of dimensions of human body. As the “site” for paper-folded clothes, human body has three levels of dimensions. The first one is of basic dimensions, and it can be deemed as combination of several spheres or polyhedrons, with different sizes at different locations, as well as difference of arcs at different locations. The second one is of wearing dimensions, human body is movable, and maximum sizes and minimum sizes for wearing and movements determine variable range of sizes of clothes. The third one is of extension dimensions, on basis of meeting requirements of aforesaid two dimensional levels, the form of human body can be re-molded and changed with clothes, and has extensible dimensions. In the end, if form control of a finished product of clothes completes three levels of human body dimensions, if molding mechanism of folding units is followed, and if setup of concepts is observed, all are considerations for evaluating quality of a product.

2.Teaching Results and Discussion

The course “spatial enclosure – the art of paper folding” obtained good teaching effect. Students were highly interested, and completed final products and drawings at a high level. Due to the existence of rational thinking and control, “replicability” has become an important feature of finished products, and it also fit well with training objective set up for the course. For the one-semester teaching of spatial basis, “spatial enclosure” continued the perception on relations between human body dimensions and space in “action-space analysis”, cultivated students with ability of rational form operations, and laid down solid foundation for practical erection of the following “art of mutually-supporting”. In the design exhibition “rationality of the art” at the end of the semester, students wearing paper-folded clothes that are designed and manufactured by themselves walked through cover structures erected by themselves, and felt the beauty of materials, the beauty of structure, and the beauty of space again. Till now, three courses have completed overall training of spatial basis.

As a classic training item at University of the Arts, “paper” and “human body” emphasize training on clothes; in this teaching plan, when “paper” and “human body” become carriers for architectural training, it emphasizes training on form composition and form logic. The exercise of “spatial enclosure” implies many basic issues of architecture, such as the issues of site, materials, construction, structure, and design. By considering human body as a site, paper-folded clothes as a building, the finished paper-folded clothes are not just independent products, but also carriers for the purpose of deliberating architectural issues. In term of architecture, a good building not only has a beautiful finished form, but also shapes the form with rationality, member units composing the form shall be rational, and splicing of members shall be rigorous. All these are covered in the process of design. Strengthening the training on thinking logic, and training design thinking in process of design will be extended into the entire process of professional training of architecture.

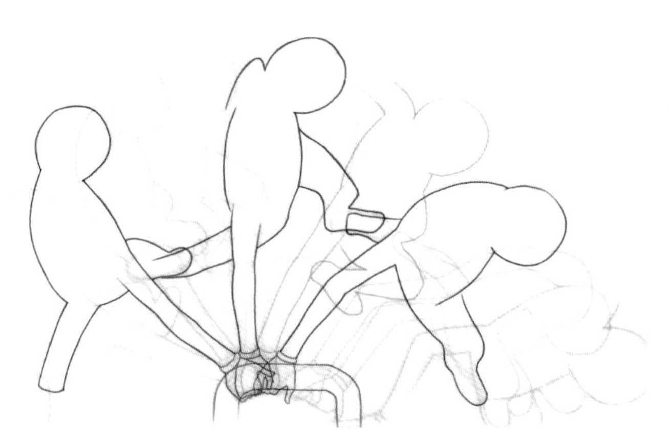

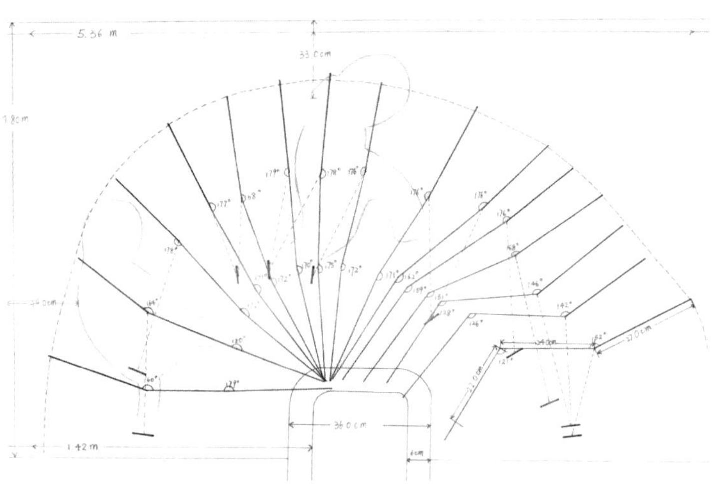
5.36 m
33.0 cm
1.42 m
36.0 cm

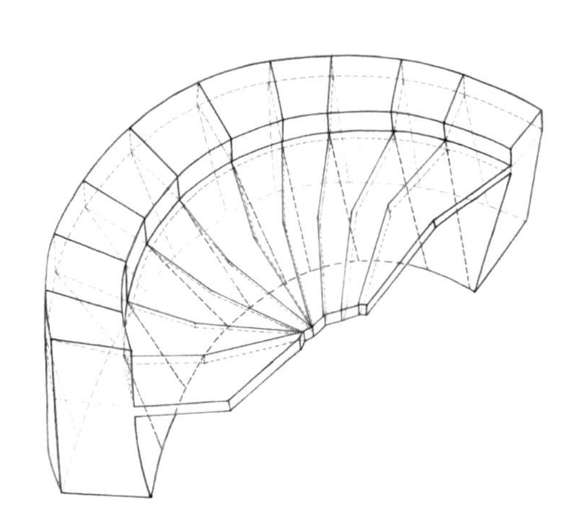

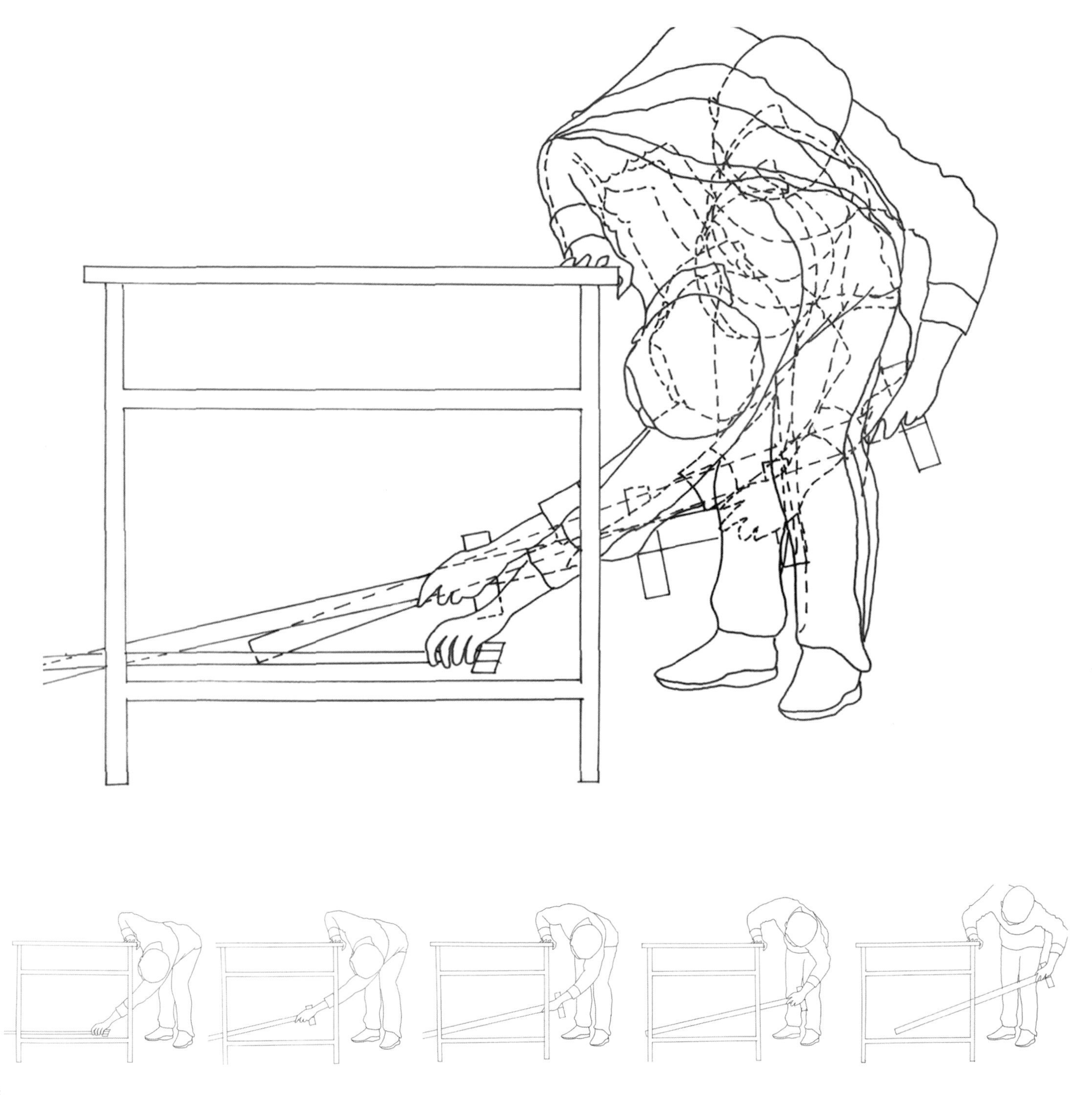

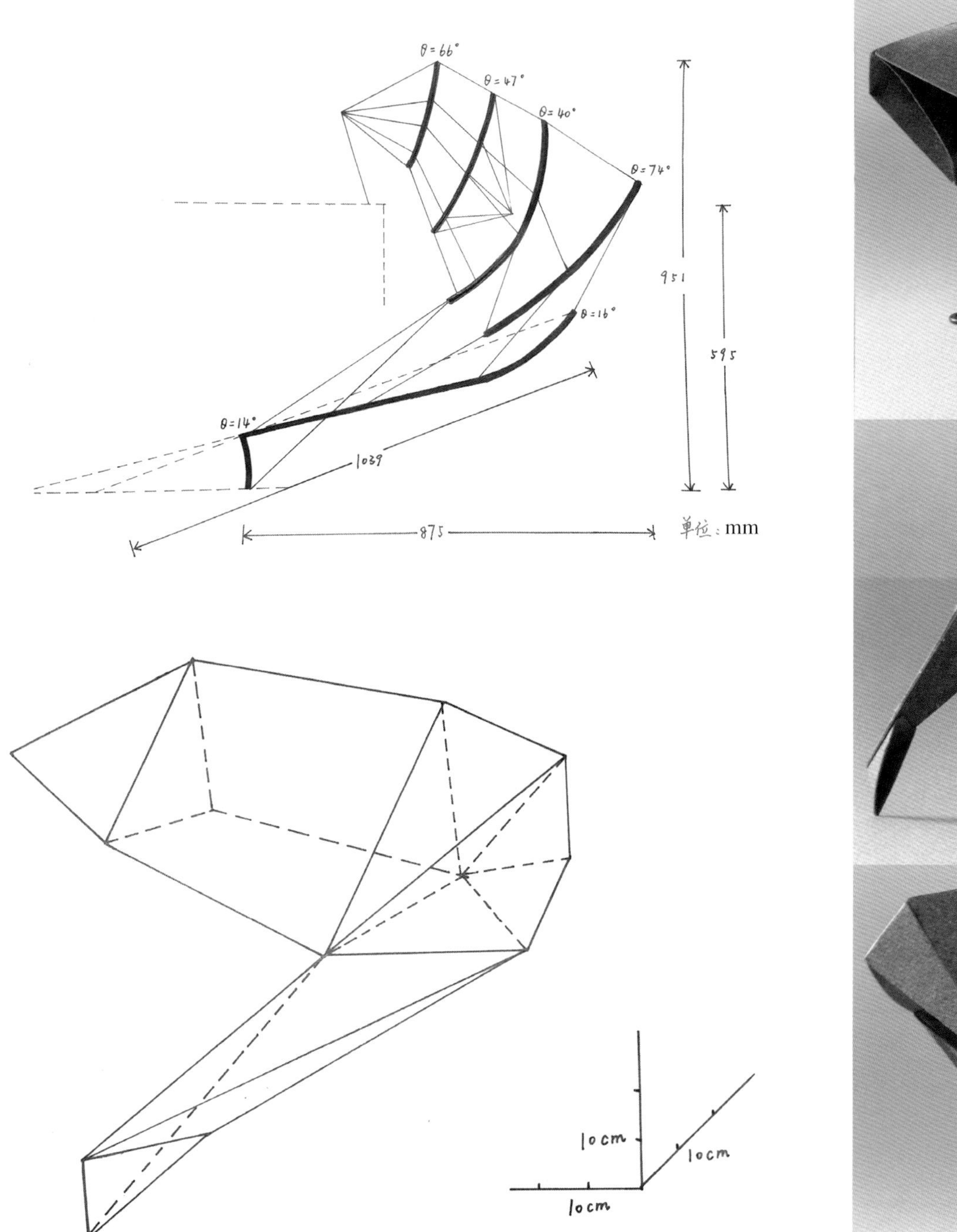
θ=66°
θ=47°
θ=40°
θ=74°
θ=16°
θ=14°
951
595
1039
875
单位：mm
10cm
10cm
10cm

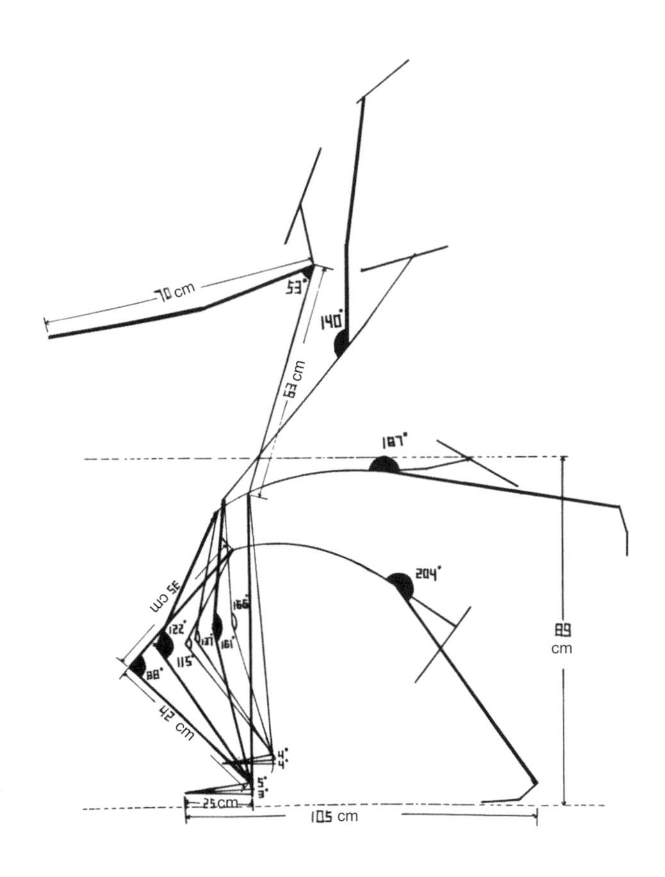

70 cm
53°
140°
63 cm
187°
204°
35 cm
166°
122°
137°
161°
115°
88°
42 cm
4°
4°
5°
3°
25 cm
105 cm
89 cm

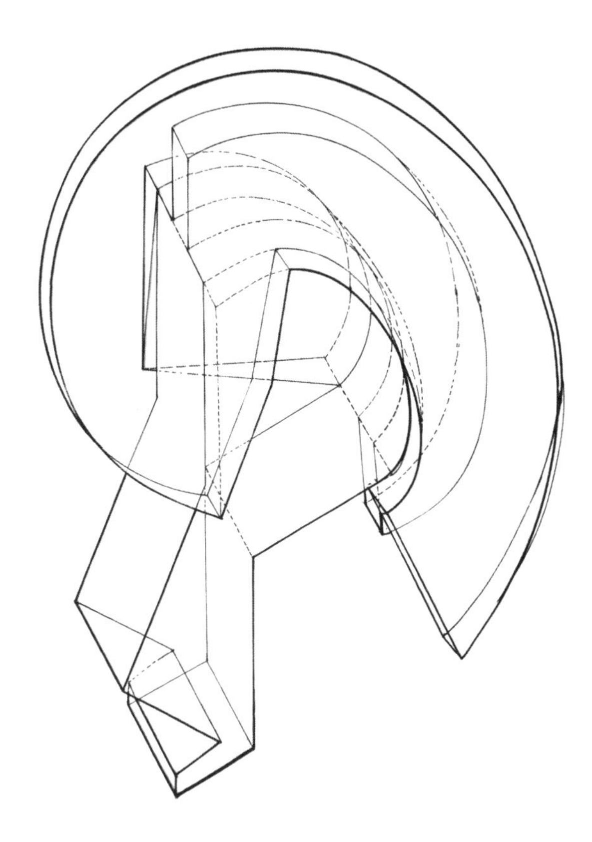

将动作、尺度抽象为几何模数与几何关系，分析动作的动态容积，针对该动作设计装置。

Make actions, dimensions into abstract geometric modulus and geometric relations, analyze dynamic volume of actions, and design device for such actions.

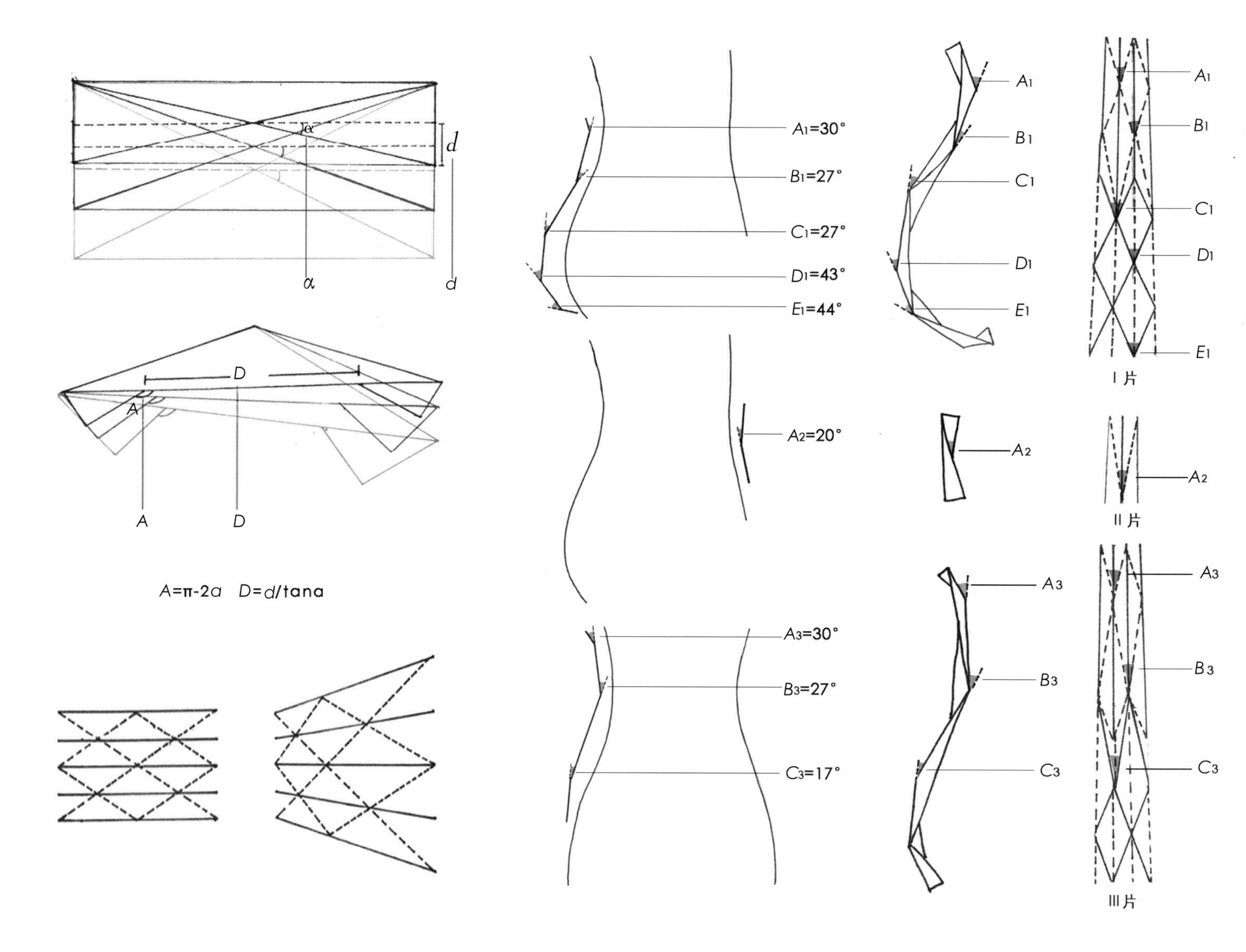

d
α
d
D
A
A
D
A=π-2α D=d/tanα
A1=30°
B1=27°
C1=27°
D1=43°
E1=44°
A2=20°
A3=30°
B3=27°
C3=17°
A1
B1
C1
D1
E1
A2
A3
B3
C3
A1
B1
C1
D1
E1
Ⅰ片
A2
Ⅱ片
A3
B3
C3
Ⅲ片

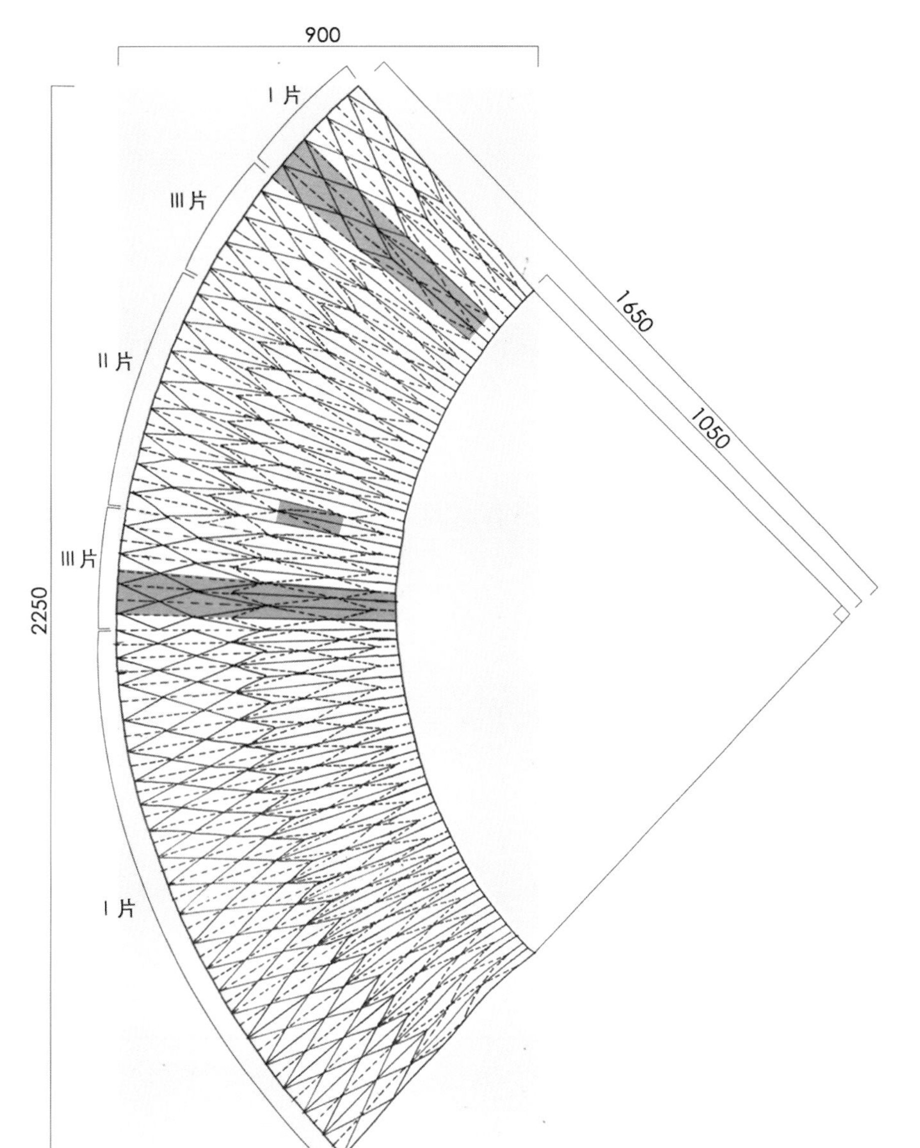

运用折纸塑形原理，包裹身体的一个部位，最终成品需满足身体尺度的三个层次，并能够改变人的形体。

Enclose one part of human body by applying principles of paper-folded molding, and a finished product shall meet three levels of human body dimensions, and can change the form of human body.

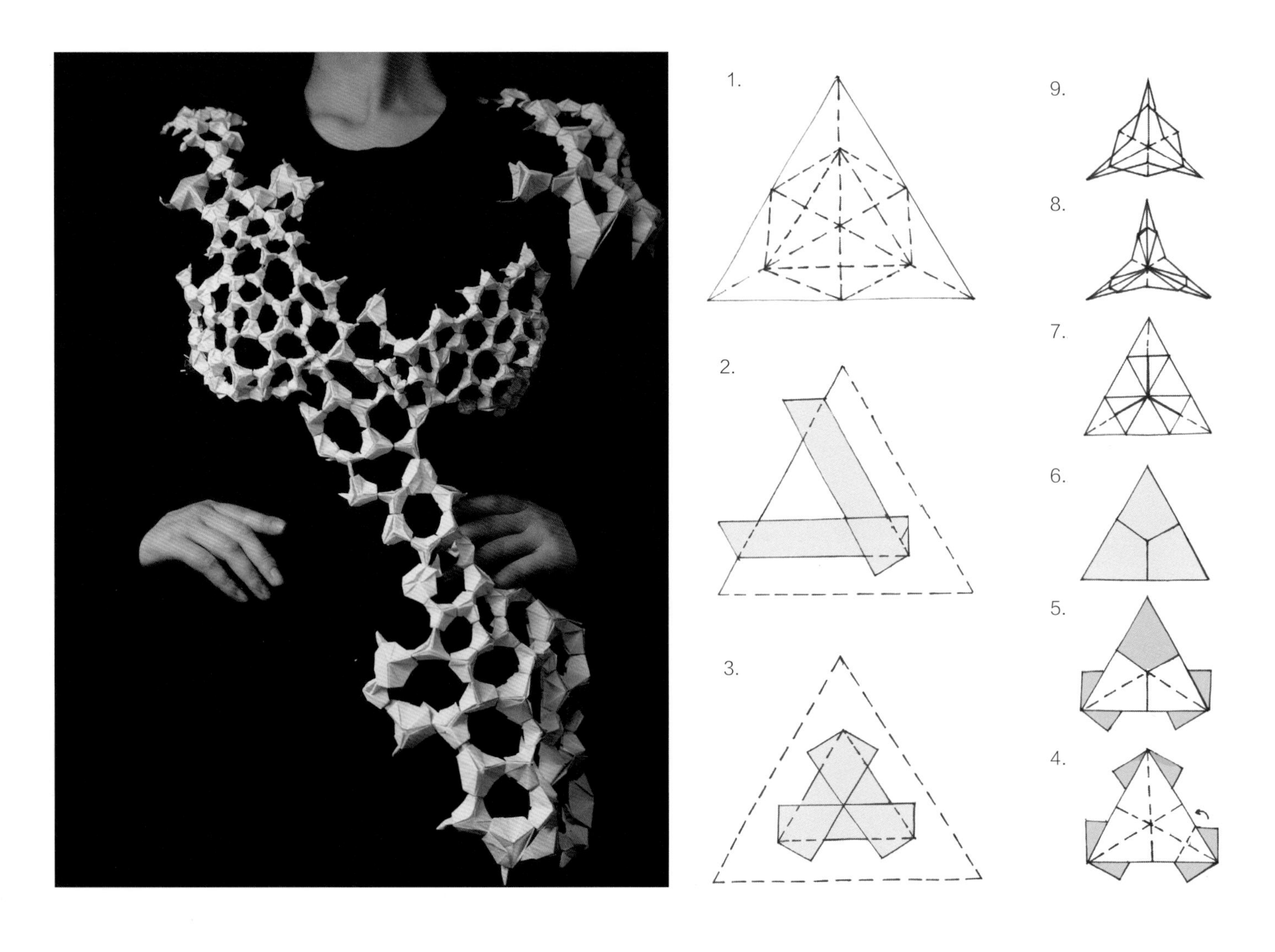
1.
2.
3.
4.
5.
6.
7.
8.
9.

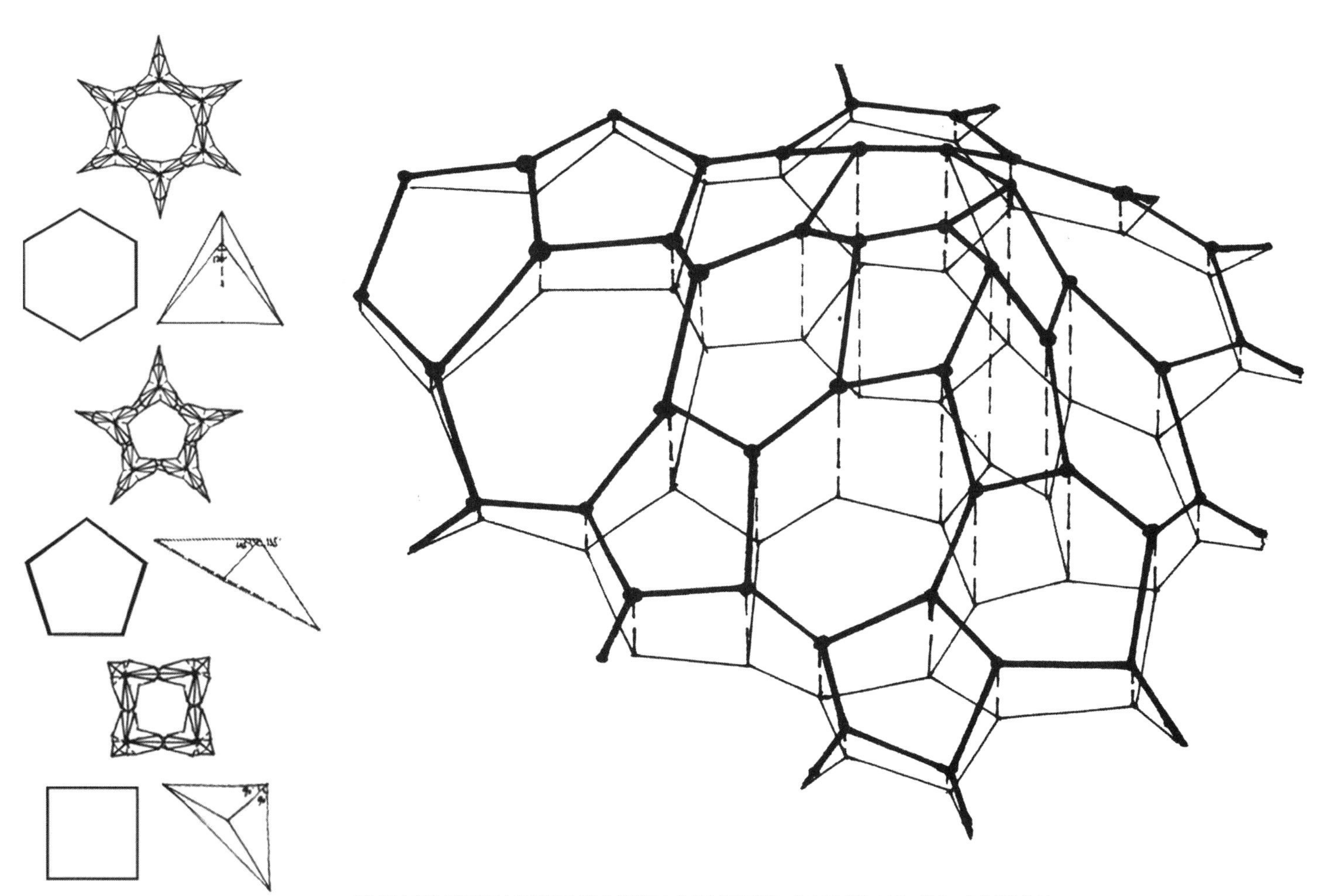

通过对白纸的折叠以及白纸折叠形成的构件单元与单元之间的拼接，体会了材料与工艺、构件、结构等的关系。

Experience relations among materials, techniques, members, and structure through folding of white paper and splicing of member units folded with paper.

repeat 4 ~ 12

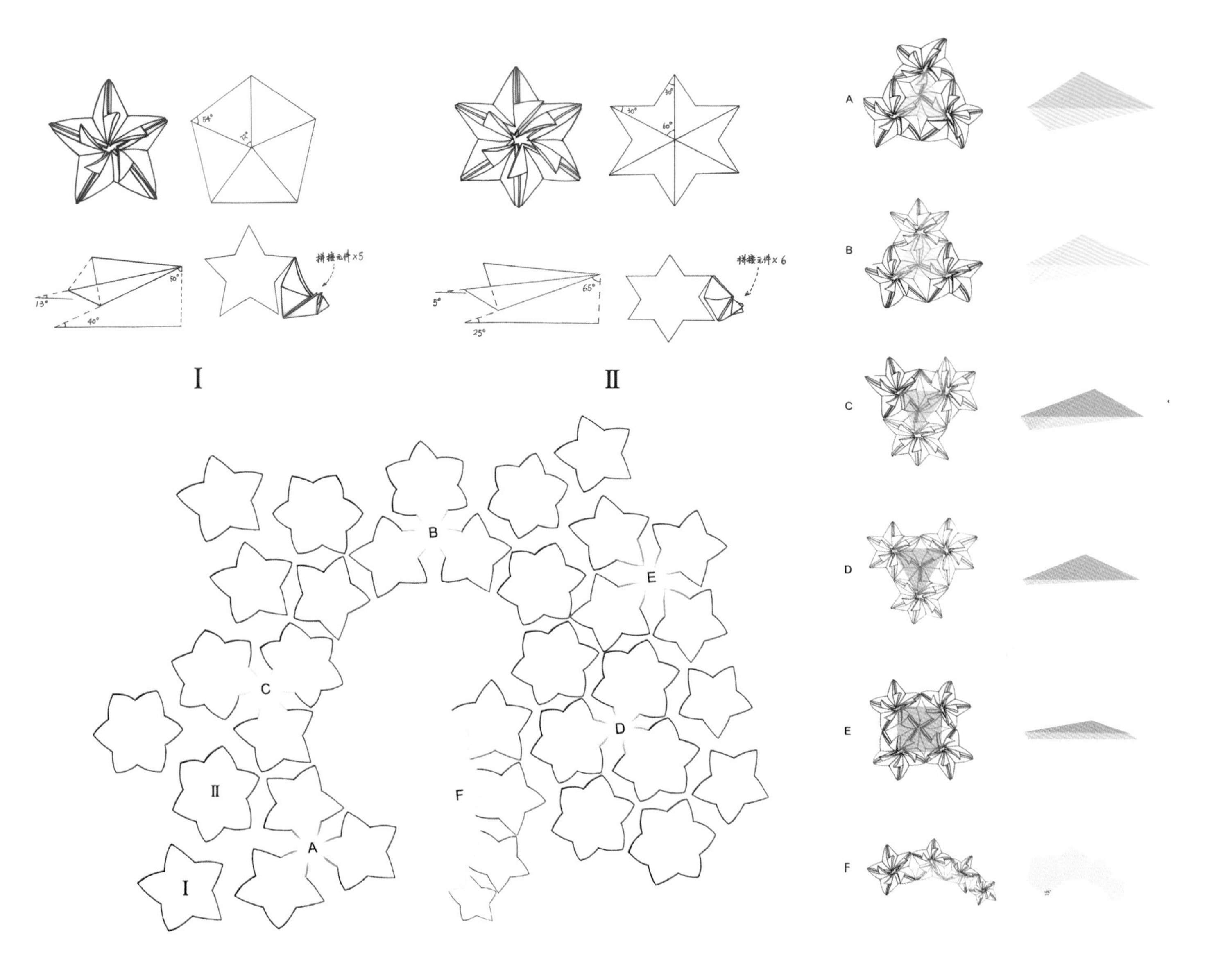
54°
72°
30°
30°
60°
13°
40°
50°
拼接元件×5
5°
25°
65°
拼接元件×6
I
II
A
B
C
D
E
F

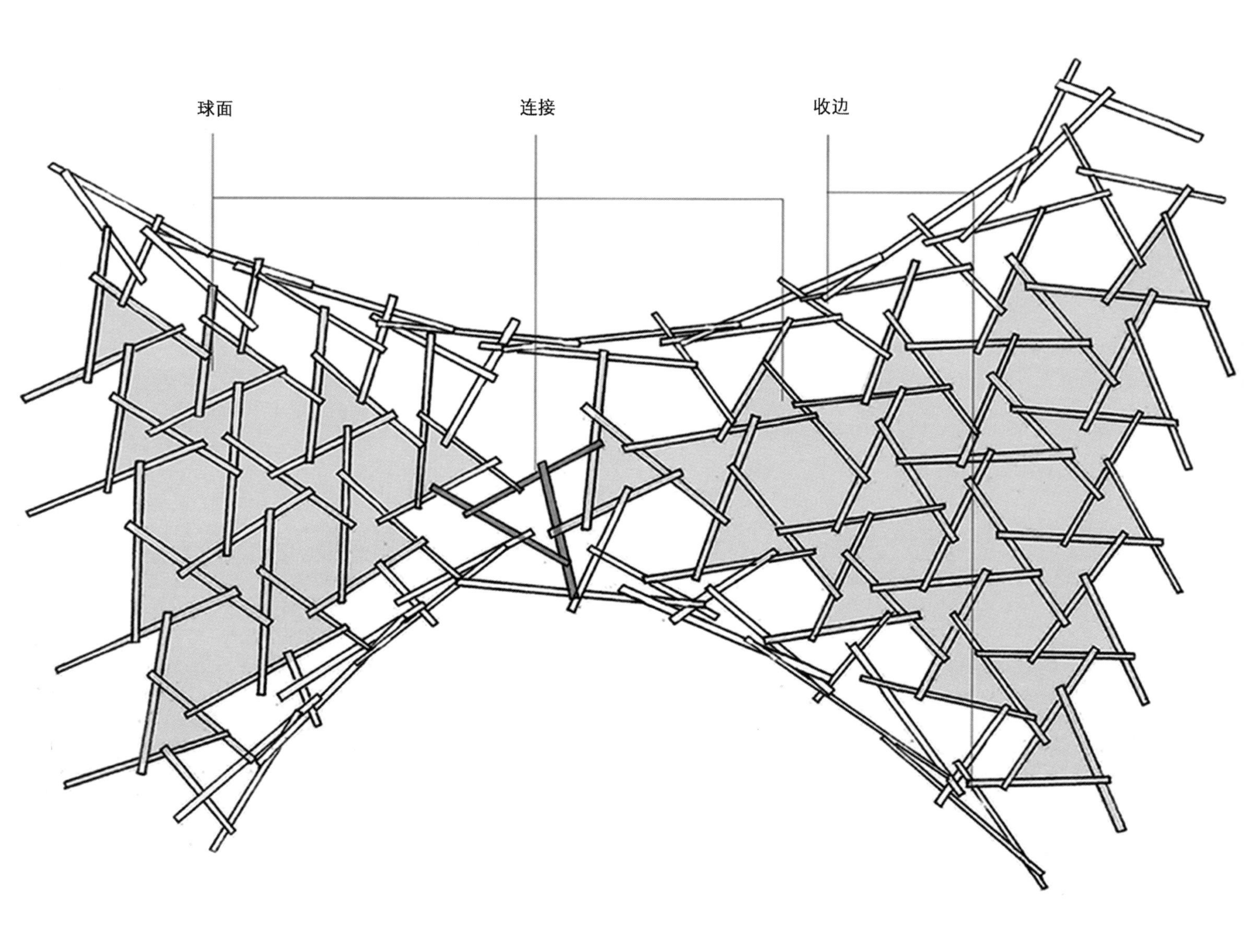
球面
连接
收边

使用 PVC 管，运用互承结构原理用小的杆件完成“大”的覆盖空间。尝试通过变化杆件的截面尺寸、搭接窗口的形状与大小、搭接方式等获得多样的空间形式。

Complete a “large” covered space with small PVC tubes by applying principles of mutually-supported structure. Try to realize various spatial forms through variations of sectional sizes of tubes, shapes and sizes of lapped windows, and ways of lap joints.

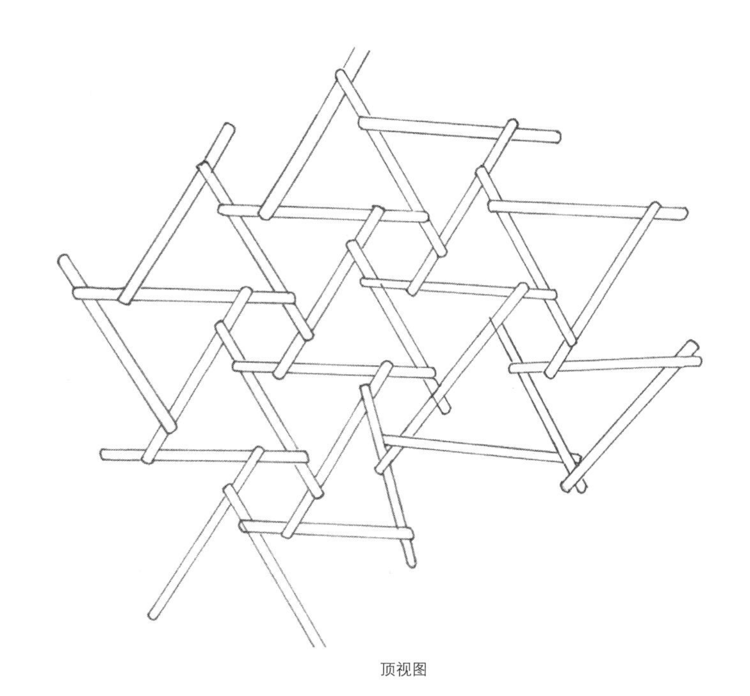

顶视图

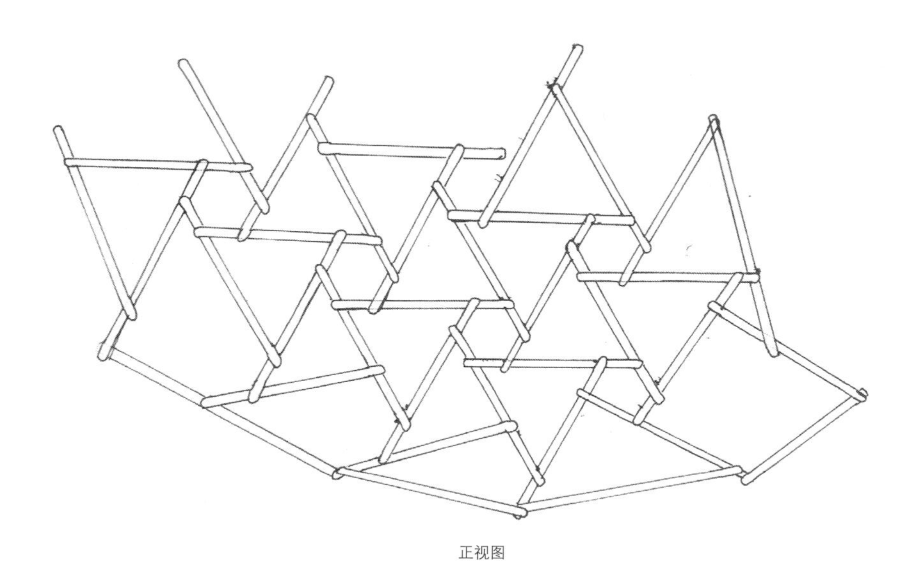

正视图

在实际搭建过程中初步建立材料、节点、造价等概念。
Preliminarily establish concepts such as materials, details, and cost in the process of practical erection.

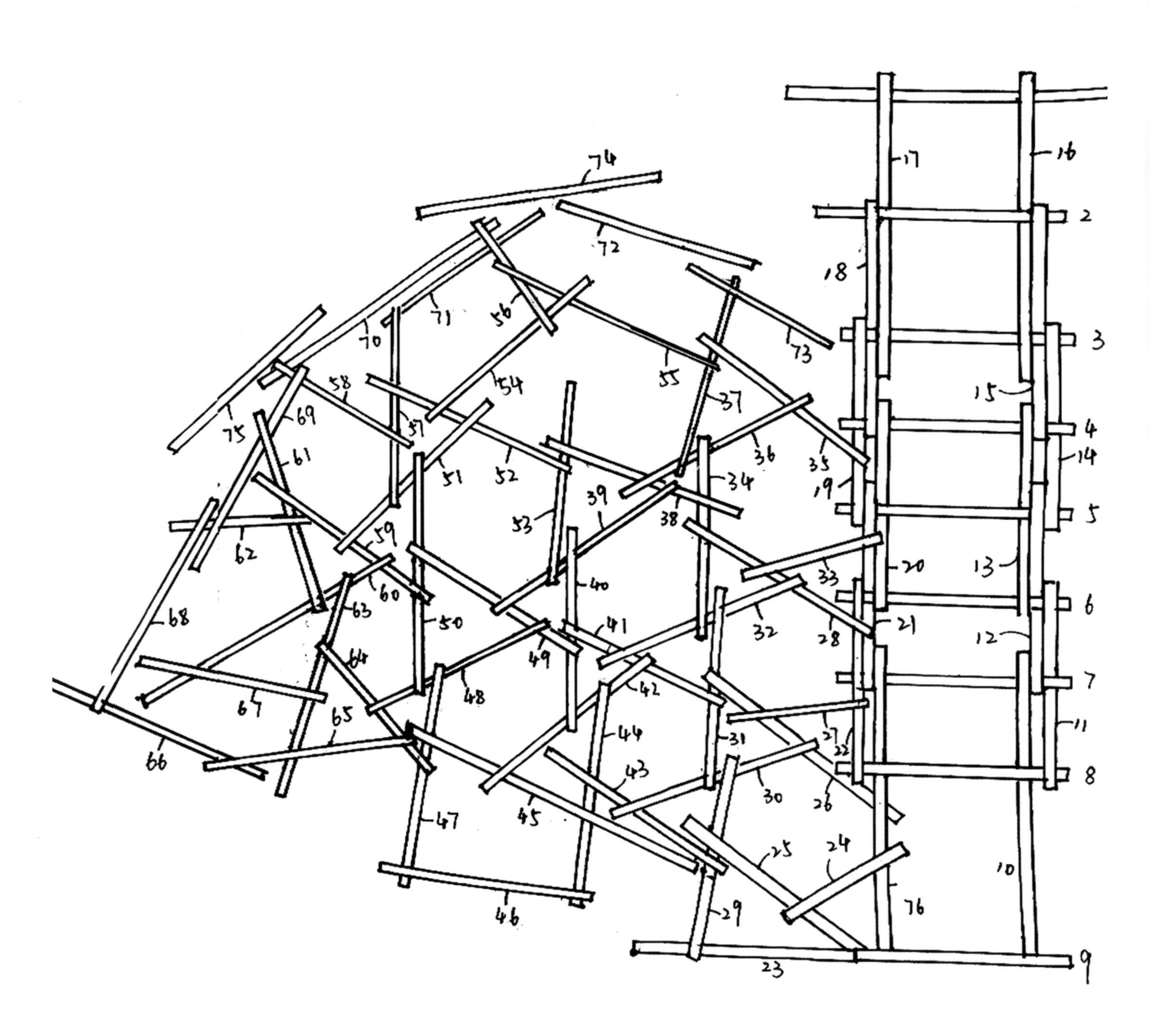

74
16
17
2
72
18
71
56
70
73
3
58
54
55
15
37
57
75
69
4
36
35
14
61
51
52
34
19
39
5
53
38
62
59
33
20
13
60
40
63
68
50
41
32
21
28
6
12
49
64
42
7
67
48
65
44
31
27
22
11
66
43
30
26
8
47
45
25
24
10
29
76
46
23
9

老城住宅设计

RESIDENCE DESIGN OF OLD TOWN

刘铨 冷天 王丹丹

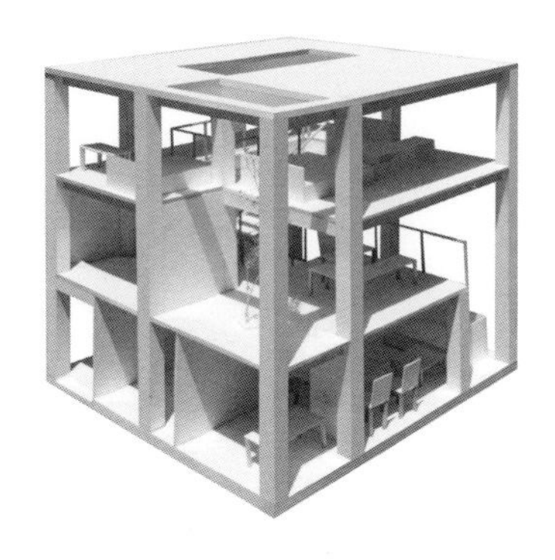

1. 课程背景

建筑设计（一）老城住宅设计课程的教学目的，是让学生综合运用建筑设计基础课程中掌握的建筑知识和表达工具，初步体验一个完整的小型建筑方案设计过程。训练的重点在于内部空间的整合性设计，同时希望学生在设计学习开始之初，能够主动去关注场地与界面、空间与功能、流线与出入口、尺度与感知等设计要素之间的紧密关系。

建筑设计的操作过程，应该围绕不同层面的建筑问题展开，并寻找解决建筑问题的最佳整合方案。建筑设计课程教案的设置，也逐渐破除了原先以建筑类型为导向的设定，转变成当今以建筑问题为导向的设定。本课程作为建筑学本科生独立完成的第一个设计任务，在尚不具备相关建筑结构、构造、技术等配套专业知识的前提下，将设计训练的重点设置为建筑空间的整合性操作之上，以期在设计教学的起步阶段，引导学生理解与体会现代建筑空间的概念和意义。

今年的教案在原有的基础上进行了调整，不再强调对水平和垂直构件的操作，以及“体积”和“片墙”两种形式语言。在使用功能上从原先的古玩店转变为私家住宅，可以更好地契合学生个人的生活经验。场地延续原先的老城历史街区，引导学生在较小的用地面积和建筑限高条件下，创造一个具备完整居家功能、流动且富有变化的内部空间，并通过合理的场地布置和流线组织，与周边的历史街区取得良好的沟通和融合。

2. 课程内容

课程设计的场地选择在学生前期进行过环境认知调研的城市历史街区中，抽出4个具有代表性的地块，并设定了相应的用地红线范围，学生选择其一完成设计。单个地块的面积在60~70m^{2}左右，单面或相邻两面临街，周边为1~2层的传统民居。建筑功能为小型家庭独立式住宅，设计者需要自行设定家庭主要成员（一对年轻夫妇、1~2位未成年子女）的职业和生活背景，并在自己的设计中加以明确体现。新建建筑面积不超过100㎡，建筑高度≤8m（指内部可用空间总高，不包括女儿墙，不设地下空间），其中应包括起居室、餐厅、主卧室、儿童房、多功能房（含工作空间和客床）、厨房、餐厅、卫生间（1~2处）以及必要的储藏空间。另外，新建建筑出入口附近应考虑至少2个以上电动自行车的停车与充电空间。

在教学进度上，本次课程共分为两个阶段。第一个阶段是典例分析（3周），通过大比例实体模型，在空间、结构、流线、尺度等方面对两个经典住宅案例进行体验认知。第二个阶段是设计操作（5周），通过专业的平立剖图纸，配合不同比例的实体模型，以及剖透视、分析图、效果图等，完成整个课程设计任务。

3. 典例分析

本课程选取了两个经典的小型独立式住宅作为案例，引导学生通过大比例实体模型（1: 50）的制作、分析性图纸的绘制，在空间、结构、流线、尺度等方面进行认知和学习。两个案例均位于日本东京的城市中心区，在场地和功能定位上同本课程的设定均有较好的契合和参考价值。

（1）团子坂住宅：由妹岛和世设计，占地面积24.3m^2，建筑面积50.4m^2。木框架结构体系，让家庭生活自身来勾勒房间的轮廓，通过标高的细微差别来区别各个空间，取消传统的隔墙界定，增添了内部空间的丰富性和趣味性。

（2）土桥住宅：由妹岛和世设计，占地面积72m^2，建筑面积60m^2。钢框架结构体系，错动的楼板开口营造出一个通高的中庭空间，为室内引入充足的光线，开放的概念在视觉上联系了每个楼层，由相互平行的楼梯进入，人们在空间中移动时会感受到结构的变化。

4. 训练重点

4.1场地与界面

建筑的形体是其内部空间的反映，而建筑的外部空间指建筑周围或建筑物之间的环境。场地从外部限定了建筑空间的生成条件，需要结合基地现状条件对场地内的建筑、道路、绿化等构成要素进行全面合理的布置，通过设计使场地中的建筑物与其他要素形成一个有机整体发挥效用。而建筑形体的外立面，作为限定城市外部空间的垂直界面，直观地反映着城市中建筑之间的关系，舒适的城市公共开敞空间也首先依赖于合理的建筑形体与布局。

4.2空间与功能

建筑设计操作最根本的目的，是获取合乎使用的空间。现代建筑日趋复杂的功能要求、建造技术和材料的突破，为建筑师创造建筑空间提供了更多可能，空间意义也成为现代建筑最重要的内涵，其意义远大于建筑内部可供使用的房间。现代建筑设计很大程度上是建筑空间的设计，建筑形式语言和设计方法均以空间作为主题展开。而建筑的功能指建筑物内外部空间应满足的实际使用要求，回答了建筑基本使用目的的问题。

4.3流线与出入口

建筑的流线俗称动线，是指人流与车流在建筑中活动的路线，根据不同的行为方式把各种空间组织起来。一方面，建筑内部各功能空间需要合理的水平、垂直交通来相互沟通与联系；另一方面，建筑的内部空间需要考虑与场地周边环境条件的合理衔接，如街道界面的连续性、出入口位置的选择与退让处理、周边建筑外墙界面（包括其上的外窗）对新建建筑的影响、建筑之间的间距与视线干扰、日照的合理使用等。

4.4尺度与感知

建筑内部的空间是供人来使用的，因此建筑中的各功能空间的尺度，都必须以人体作为基本的参照和考量，并结合人体的各种行为活动方式，来确定合理的建筑空间尺寸。在空间形式处理中注意通过图示表达理解空间构成要素与人的空间体验之间的关系，主要包括尺度感和围合感。建筑尺度还受到建造条件的限制，并与环境存在参照的关系。

5. 时间安排

第一周：收集典例相关资料，制作工作模型研究空间和结构关系。

第二周：深入绘制分析图纸，制作大比例实体模型（带内部家具）。

第三周：完成典例分析的图纸绘制和实体模型制作。

第四周：场地调研，制作场地模型，构思初步方案。

第五周：确定基本设计方案并细化推敲各设计细节。

第六周：深化1:50图纸，开始照片拼贴效果图的制作。

第七周：制作1:20剖透视和分析图，制作1:20大比例模型。

第八周：整理图纸、排版并完成课程答辩。

6. 教学体会

在教案的设定和限制下，绝大部分同学都围绕训练重点完成了一次整合性的建筑空间组织设计。错动的楼板、不同位置的开孔等设计手法，通过框架式结构整合在一起，形成了流动且富有变化的内部空间。大比例的实体模型，配合细致的剖透视分析图，较好地将内部的空间组织效果直观地展示出来，最终的设计成果达到了教案预期的训练要求。不过，国内现行的某些行业规范，尤其是垂直交通（楼梯、踏步）上的限制，在面积指标的分配上对内部空间的灵活分配带来了较大的制约，需要在未来的教案设定上加以调整。

I.Background of the Course

The teaching objective of the course of Old South Residential Design of Architectural Design (I) is allowing students to preliminarily experience the design process of a complete small building design through comprehensive application of architectural knowledge and expression tools mastered in the course of Fundamentals of Architectural Design. Key point of the training is the integrated design of interior spaces, and it expects students to pay attention to the close relations between design elements such as site and interface, space and function, circulation and doorway, dimensions and perception at beginning of the design learning process.

The operation process of architectural design should be unfolded by centering on different levels of architectural issues, and seek an optimal integration plan to solve architectural issues. In term of setup of teaching plan of architectural design course, it gradually got rid of the traditional building type-oriented setup, and has been transformed into the architectural issue-oriented setup of today. As the first design task to be completed independently by undergraduate students of architecture, and given that they still do not have associated professional knowledge about architectural structure, construction, and technologies, the focus of design training of this course is the integrated operation of architectural space, so as to guide students to understand and experience the concept and meaning of modern architectural space at beginning of design teaching.

The teaching plan for this year has some adjustment on existing basis, and no longer emphasizes operation of horizontal and vertical members, as well as the two types of language of “volume” and “individual wall”. In term of function, it is changed from an antique shop to a private residence, which can better match with personal life experience of students. The site is still located in an historic urban district, and it is intended to guide students to create an interior space that has complete living functions and is circulated and rich of variations, with a small land area and under building height limit, and to realize good communication and integration with surrounding historical urban district through reasonable site layout and circulation organization.

2.Content of the Course

The site selected for course design is located within the historical urban district where students have done environmental cognition investigation in earlier stage, four representative land parcels are chosen, and relevant land boundary lines are determined, and students may choose one of them to complete their design. Area of individual land parcel is about 60~70m^2, one side faces or two adjacent sides face the street, and around land parcels are one or two-floor traditional folk dwellings. Architectural function is a small household detached house, the designer shall determine occupation and life background of primary family members (a young couple, 1 or 2 minor children) by himself / herself, and reflect it expressively in his/her design. New construction area shall not exceed 100m^2, building height ≤ 8m (it refers to height of interior available space, and does not include parapets, and no basement), it shall include living room, dining room, master bedroom, children’s room, multi-functional room (including work space and guest bed), kitchen, bathrooms(1~2) , as well as necessary storage space. Furthermore, parking and charging space for at least two electric bicycles shall be considered close to doorway of the new building.

Teaching progress of this course consists of two phases. The first phase is for typical analysis (3 weeks), and students will experience and perceive two classic residential cases in aspects of space, structure, circulation, and dimensions with two large-scale physical models. The second phase is for design operation (5 weeks), and students will complete the entire course design task with professional plans, elevations, and profiles, together with physical models of difference scales, as well as sectional perspectives, analysis drawings, and renderings.

3.Typical Analysis

This courses selected two cases of classic small detached house to guide students to realize perception and learning in aspects of space, structure, circulation, and dimensions through establishment of large-scale physical model (1:50), and preparation of analytic drawings. Both cases are located at central urban area of Tokyo, Japan, which fit well with and has good reference value for setup of this course in terms of site and functional positioning.

(1) Dangozaka House: it is designed by Kazuyo Sejima, covers a land area of 24.3m^2 floor area of 50.4m^2. It has a wooden-frame structure system, allows family life itself to draw profile of rooms, and it differentiates different spaces with slight difference between elevations, which eliminates traditional boundary of partitions, and adds richness and enjoyment of interior space.

(2)Tsuchihashi House: it is designed by Kazuyo Sejima, covers a land area of 72m^2, floor area of 60m^2. It has a steel frame structure system, staggered openings in floor slab create a full-height atrium space, bring in sufficient light into the house, the concept of open space links all floors visually, they can be accessed with stairs that are parallel to each other, and the person who moves in the space can feel change of the structure.

4.Key Points of Training

4.1Site and Interface

The building form is reflected with its interior space, while exterior space of the building refers to the environment around the building or between buildings. The site defines the generation conditions of architectural space from outside, it requires carrying out complete and reasonable layout of elements such as buildings, roads, and landscaping within the site in combination with current conditions of the base, and shaping buildings and other elements within the site into an organic integral to realize their functions through design. And the facade of the architectural form, as vertical interface defining external urban space, reflects directly relations between the city and buildings, and comfortable public open space in the city also relies on reasonable architectural forms and layout in the first place.

4.2Space and Function

The fundamental target of architectural design operation is obtaining available and suitable spaces. Increasingly complicated functional requirements, and breakthroughs of construction technologies and materials of modern architecture offer architects much more possibility to create architectural space, and the meaning of space has become the most important implication of modern architecture, such implication is more important than available rooms within a building. Modern architectural design is the design of architectural space to a great extent, the language and design methods of architectural forms are unfolded centering on the theme of space. And functions of a building refer to actual use requirements that shall be satisfied by interior and exterior space of a building, which answer the question of basic use purpose of buildings.

4.3Circulation and Doorway

Circulation of building is also known as flow lines, meaning routes of activities of flows of human and vehicles within a building, and various spaces are organized on basis of difference behavior modes. On the one hand, various functional spaces within the building require mutual communication and linking with reasonable horizontal and vertical traffic; and on the other hand, interior space of a building shall take into account the reasonable connection with environmental conditions around the site, for example, continuity of street interface, location selection and setback treatment of doorways, effect of exterior wall interface of surrounding buildings (including their exterior windows) on the new building, spacing and sight interference between buildings, and reasonable utilization of sunlight, etc.

4.4Dimensions and Perception

Spaces within a building are to be used by human, so dimensions of various functional spaces within the building must make human body as basic reference and consideration, and determine reasonable dimensions of architectural spaces in combination with various behaviors and activity modes of human body. In the process of spatial form treatment, attention shall be paid to understanding the relations between elements of space and human experience on space with graphic expression, including sense of dimensions and sense of enclosure. Building dimensions are also limited by construction conditions, and have referential relationship with environment.

5.Schedule

Week 1: Collect data about the classic case, and make work models to study relations between space and structure.

Week 2: Prepare in-depth analysis drawings, and make large-scale physical mode (with furniture inside).

Week 3: Complete analysis drawings for the classic case and construction of physical model.

Week 4: Carry out site investigation, make site models, and conceive preliminary plan.

Week 5: Determine basic design plan, and deliberate and refine design details.

Week 6: Prepare 1:50 detailed drawings, and start to splice renderings with photos.

Week 7: Prepare 1:20 sectional perspectives and analysis drawings, and make 1:20 large-scale model.

Week 8: Compile drawings, compose and complete course defense.

6.Teaching Experience

Under the setup and limits of teaching plan, most students completed one integrated design of architectural space organization by centering on key training points. Design methods such as staggered floor slabs and openings at different locations were integrated together through a framed structure, shaping fluid and varied interior spaces. Large-scale physical mode, and carefully coordinated sectional perspective analysis drawings presented effect of interior space organization in a favorable and intuitive way, and consequently the design requirements satisfied expected training requirements in the teaching plan. However, some current industrial standards in China, especially the limits on vertical traffic (stairs, treads), have significant restriction on flexible allocation of interior space in term of distribution of area indicators, which shall be adjusted in setup of future teaching plan.

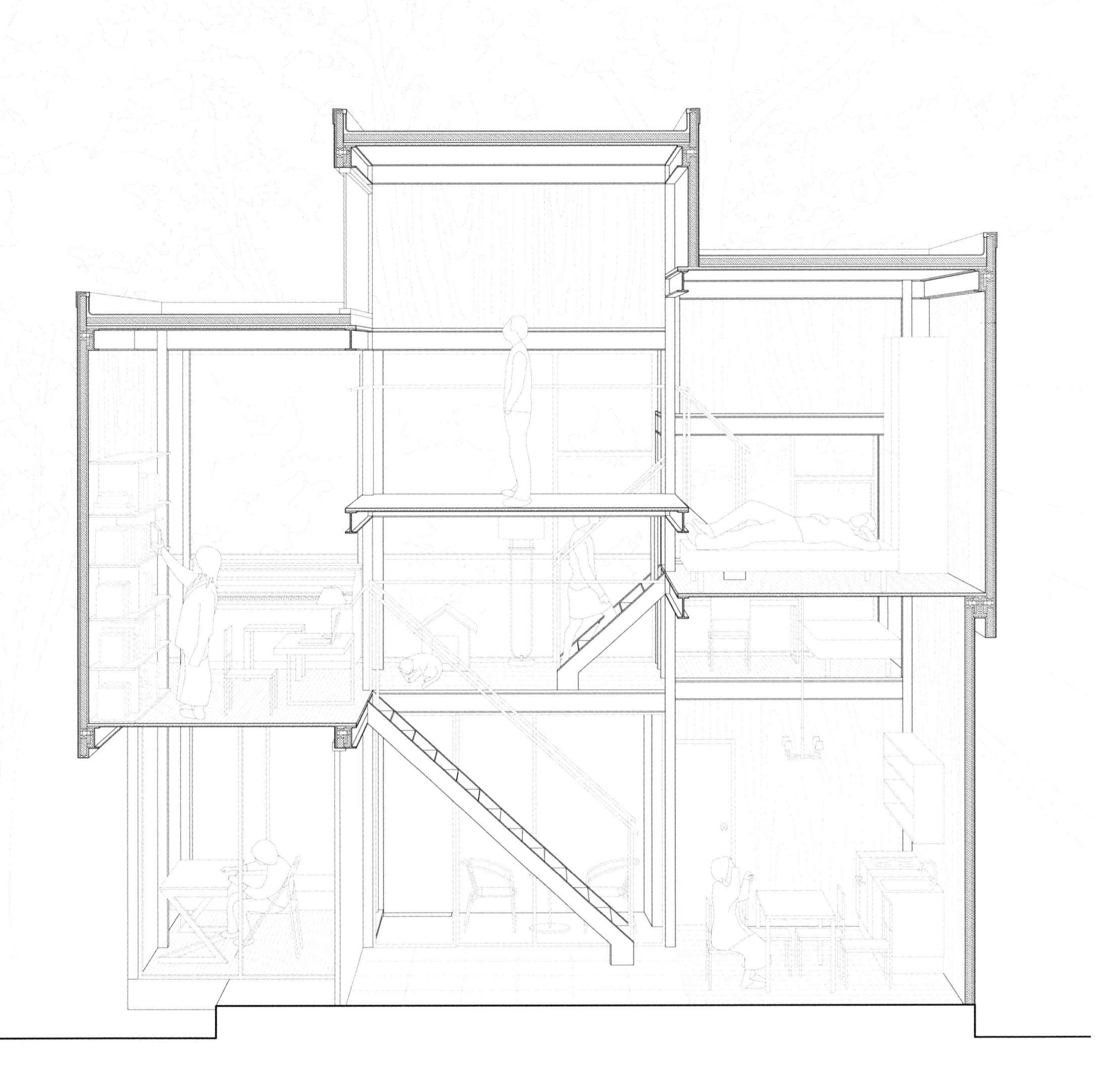

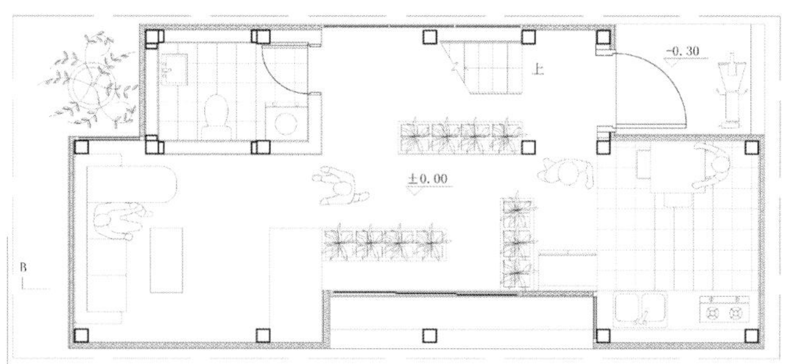

-0. 30
上
±0. 00
B

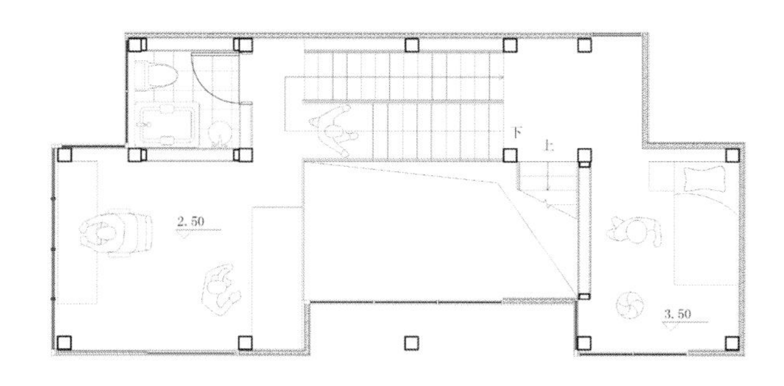

下
上
2. 50
3. 50

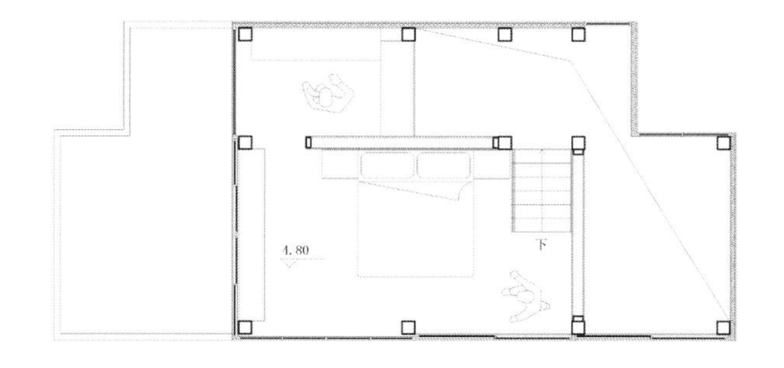

4. 80
下

在较小的用地面积和建筑限高条件下，创造一个具备完整居家功能、流动且富有变化的内部空间。

One interior space with complete living functions, fluidity and rich of variation was created with a relative small land area and under condition of building height limit.

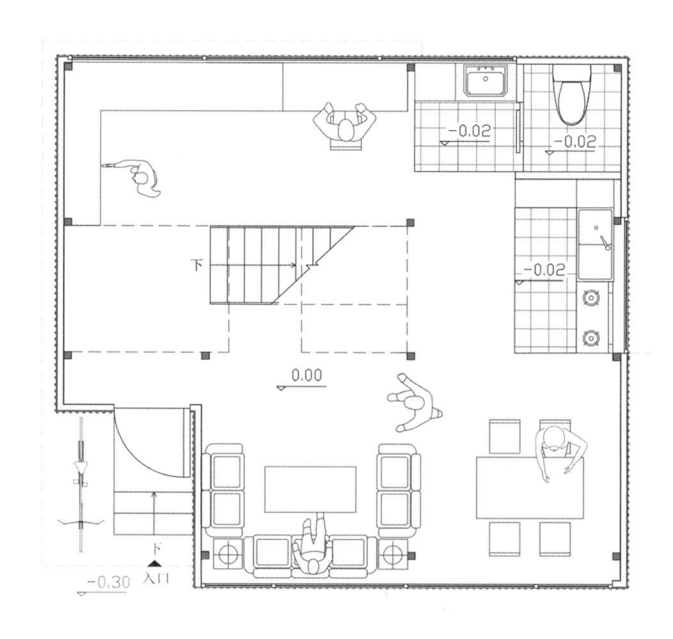

-0.02
-0.02
下
-0.02
0.00
下
-0.30
入口

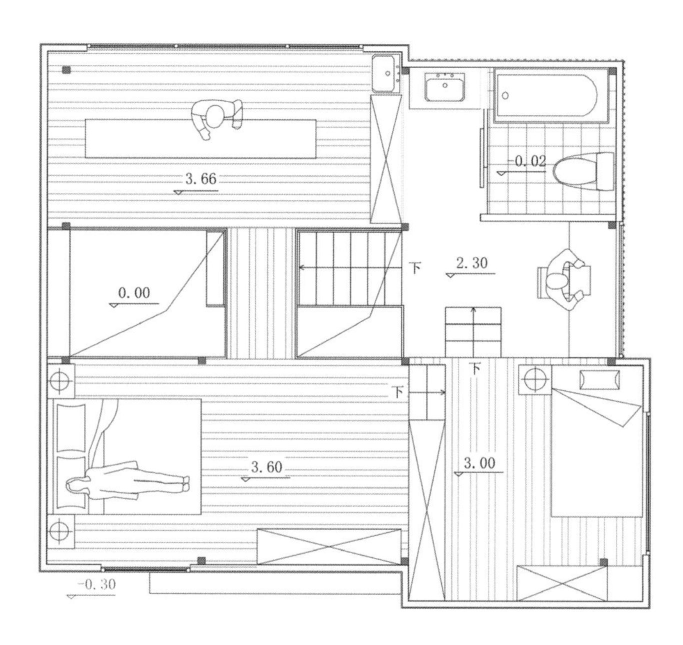

3. 66
0. 02
0. 00
下
2. 30
下
下
3. 60
3. 00
-0. 30

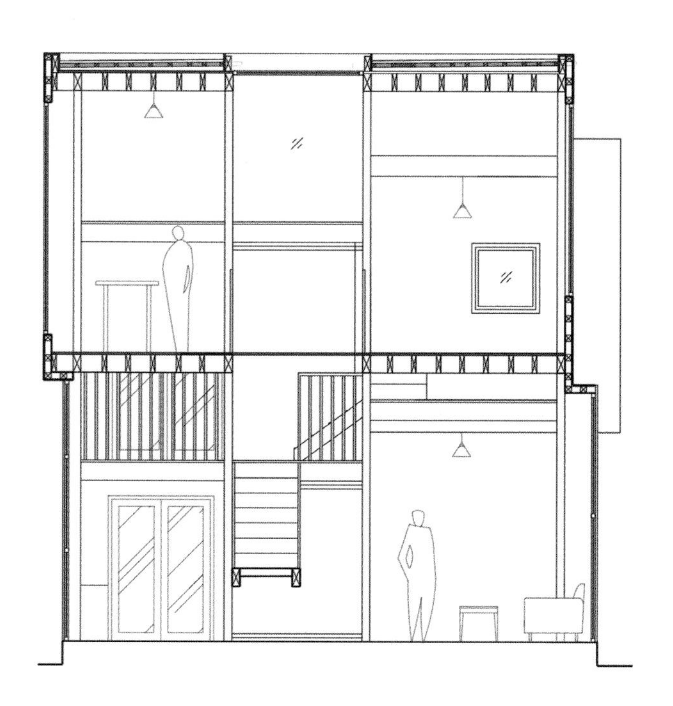

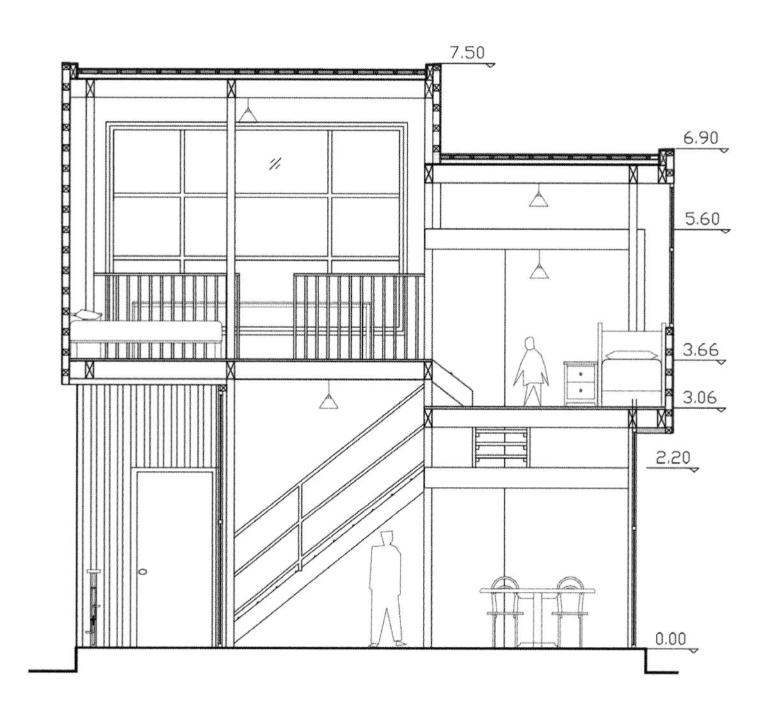

7.50
6.90
5.60
3.66
3.06
2.20
0.00

错动的楼板、不同位置的开孔等设计手法，通过框架式结构整合在一起，形成了流动且富有变化的内部空间。

Design methods such as staggered floor slabs and openings at different locations were integrated together through a framed structure, shaping fluid and varied interior spaces.

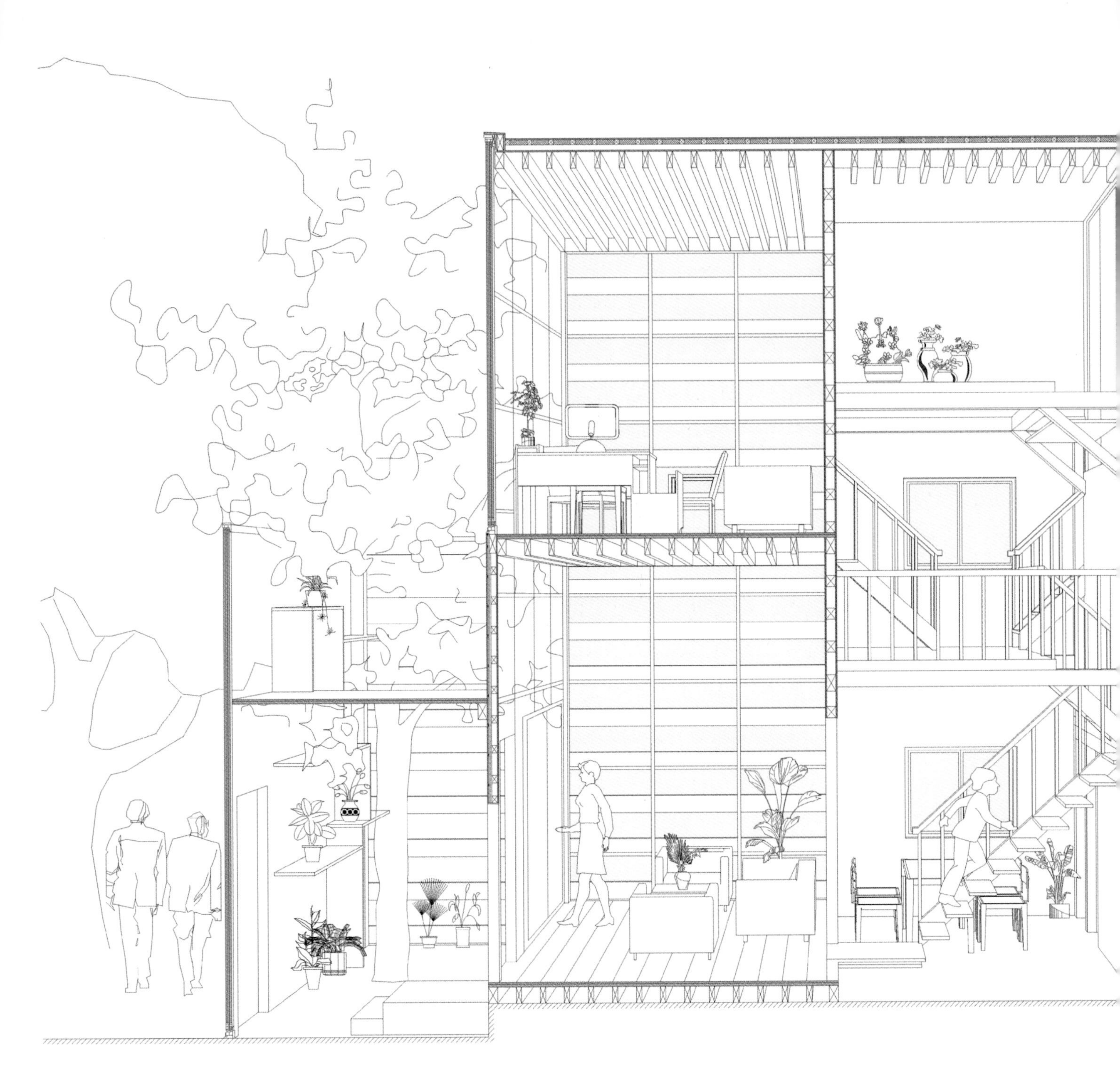

通过合理的场地布置和流线组织，与周边的历史街区取得良好的沟通和融合。
Good communication and integration with surrounding historic urban district were obtained through reasonable site layout and circulation organization.

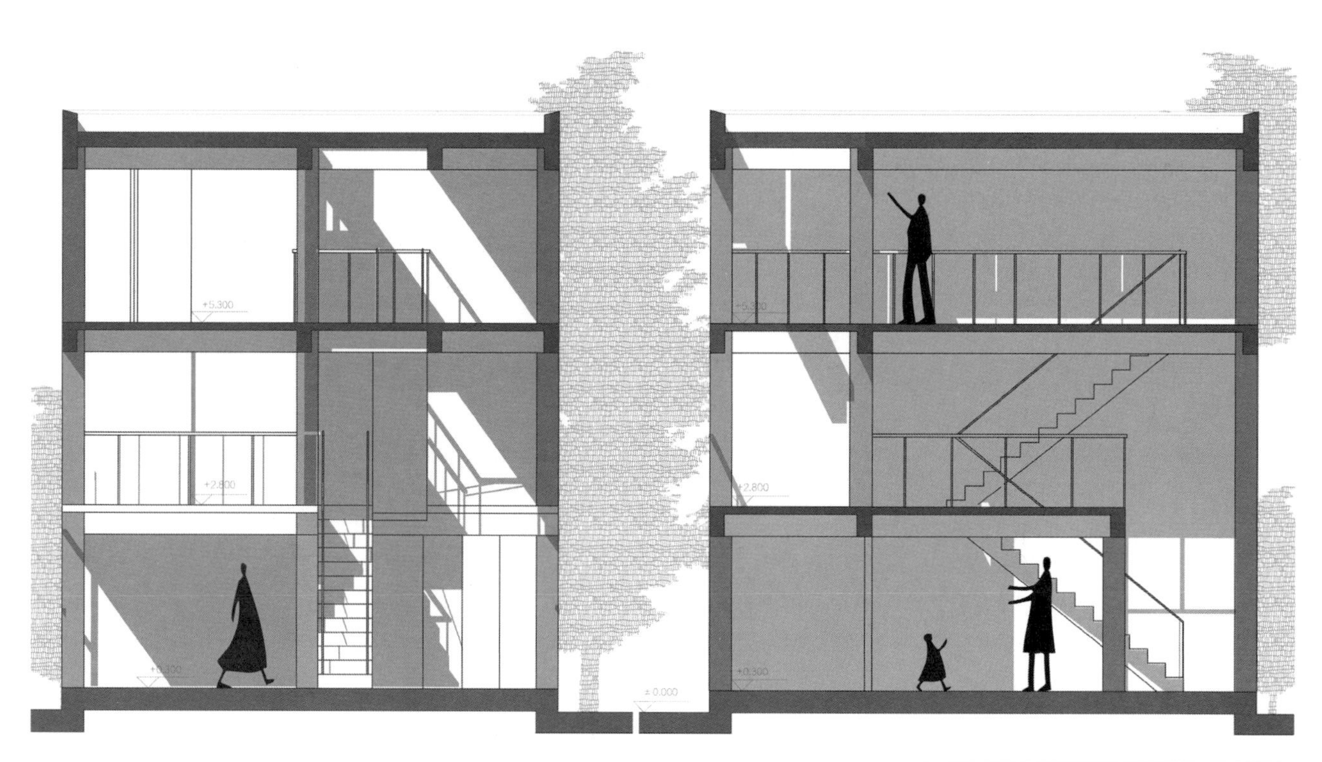
+5.300
+2.800
+0.300
± 0.000

傅抱石美术馆设计

DESIGN OF FU BAOSHI GALLERY

周凌 童滋雨 窦平平

设计课程围绕“空间”(space) 和“流线”(circulation) 两个教学主题，学习建筑空间组织的技巧和方法，训练空间的效果与表达。空间问题是建筑学的基本问题，因而课题基于复杂空间组织的训练和学习。从空间秩序入手，安排大空间与小空间、独立空间与重复空间，区分公共与私密空间、服务与被服务空间、开放与封闭空间。同时，空间的串联形成序列，需要有效组织流线，并且充分考虑人在空间中的行为和空间感受。以模型为手段，辅助推敲。设计阶段分体积、空间、结构、围合等，最终形成一个完整的设计。

另外，题目选择加入了特定环境和特定主题，课程要求研究中国传统空间类型，提出一个空间原型，在此基础上发展设计方案。课程开始，除了基地分析，还要求学生对设计对象：傅抱石的绘画以及中国绘画艺术，进行了研究和分组讨论。

基地位于南京汉口西路 132 号傅抱石生前旧居东侧，南侧为其执教的南京师范大学（前金陵女子大学）校园。目前，由于展馆规模较小，纪念馆需要进行扩建。建筑退让需要满足《南京市城市规划管理条例》的要求。本次设计需针对傅抱石作品进行专门设计。设计包含四大功能：收藏、展示、研究、公民素质教育与对外文化交流，训练处理空间收放、空间序列、服务流线、大小空间安排等综合能力，还要面对复杂的城市环境、山地高差、历史建筑、保留树木的关系，是一个综合性较强的课程设计。

成果分为四部分：①空间与环境：总平面（1:500）；序列人眼透视（环境融入）。②空间基本表达：平立剖面（1:200）。③空间解析与表现：轴测分析图；水平楼板序列轴测；垂直墙体序列轴测；仰视轴测；剖透视；人眼透视。④手工模型：1:500 总图体量模型；1:200/300 带环境模型。

8 周时间，教学环节分为基地、空间原型、文化、功能几方面的重点训练，最后完成设计。大多数同学能够按照教学进度完成设定目标，教学效果良好，几位同学提出了创造性方案。

By centering on two teaching topics of “Space” and “Circulation”, this design course is aimed for learning techniques and methods of architectural space organization, and training on effect and presentation of space. Space is a basic issue for architecture, and the course is based on training and study on organization of complex spaces. Start from spatial order to arrange large and small spaces, independent space and overlapped space, and to distinguish public and private spaces, serving and served spaces, open and closed spaces. Meanwhile, linking spaces to shape sequence requires effective organizational circulation, as well as full consideration of behaviors, spatial feeling of people in space. Use models as means to assist deliberation. Design stages include volume, space, structure, and enclosure, and shape a complete design in the end.

In addition, specific environment and special topics are selected and added to the subject, and the course requires studying on traditional space types in China, coming up with a prototype of space, and developing design plan on basis of it. At beginning of the course, besides analysis on the base, it also requires students to study and discuss in groups the paintings of Fu Baoshi and Chinese painting art.

The site is located at east side of the former residence of Fu Baoshi where he lived before he passed away, 132 West Hankou Road, Nanjing, and at south side there is the campus of Nanjing Normal University. Today, the Memorial Hall is to be expanded given the small area of its exhibition hall. And setback of the building shall satisfy requirements in the Administrative Regulations on Urban Planning of Nanjing. Special design for works of Fu Baoshi is required in this design course.The Art Gallery undertakes four major functions: collection, display, research, citizen quality education and cultural exchange with foreign countries. The training is carried out for comprehensive ability of handling space deploying and retracting, space sequence, service circulation, and arrangement of large and small spaces. It also needs to face relations among complex urban environment, height difference of hill land, historical building, and trees preservation. It is a course design with strong comprehensive nature.

The outcome consists of four parts: ① Space and environment: master plan (1: 500); human-eye perspective of sequence (blending in of environment). ② Basic presentation of space: plans, elevations, and profiles (1: 200). ③ Analysis and expression of space: axonometric analysis drawing; axonometric view of horizontal floor sequence; axonometric view of vertical wall sequence; upward axonometric view; sectional perspective; human-eye perspective . ④ Hand-made model: 1: 500 massing model of master plan; 1: 200/300 model with environment.

The course spans 8 weeks, its teaching links include important training in aspects of site, space prototype, culture, and function, and complete the design in the end. Most students could complete set objectives according to the teaching progress, with good teaching effect, and some students presented plans with creativity.

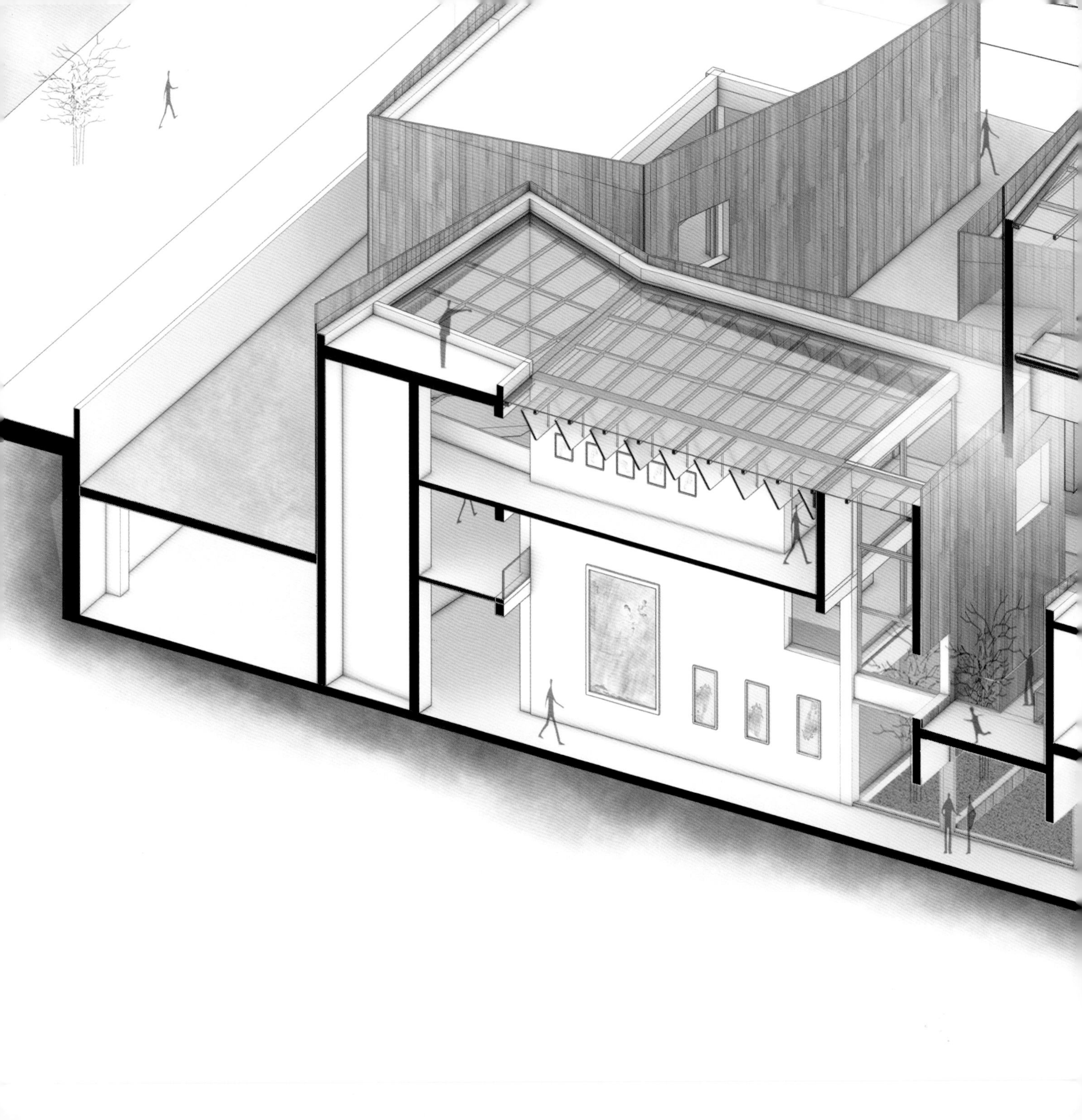

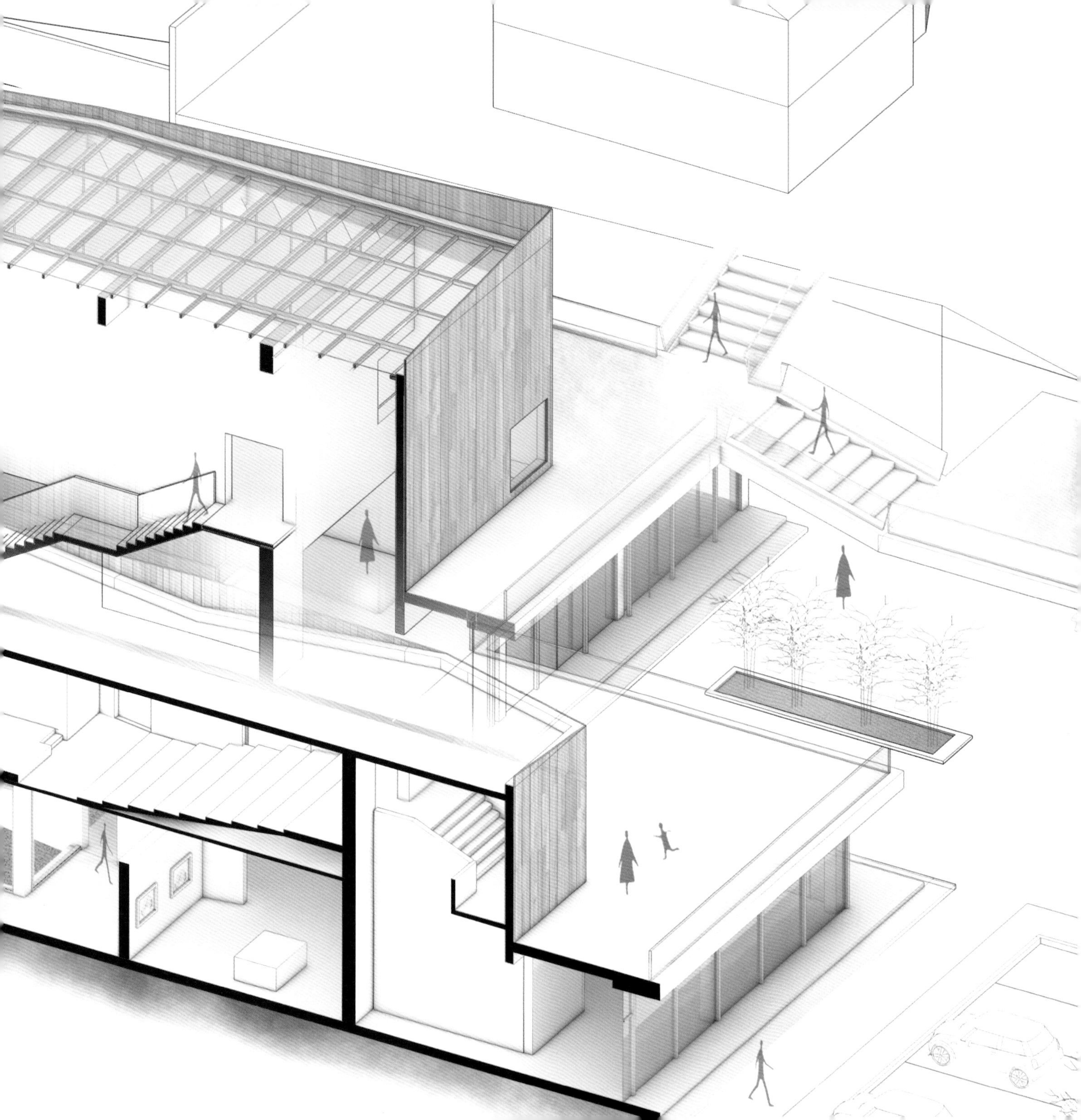

在后部引入公共空间，通过一系列高差及垂直界面的设置，为不同形式的公共活动提供空间。

Bring in public space at the back side and provide space for different types of public activities through a series of dispersion and vertical interface.

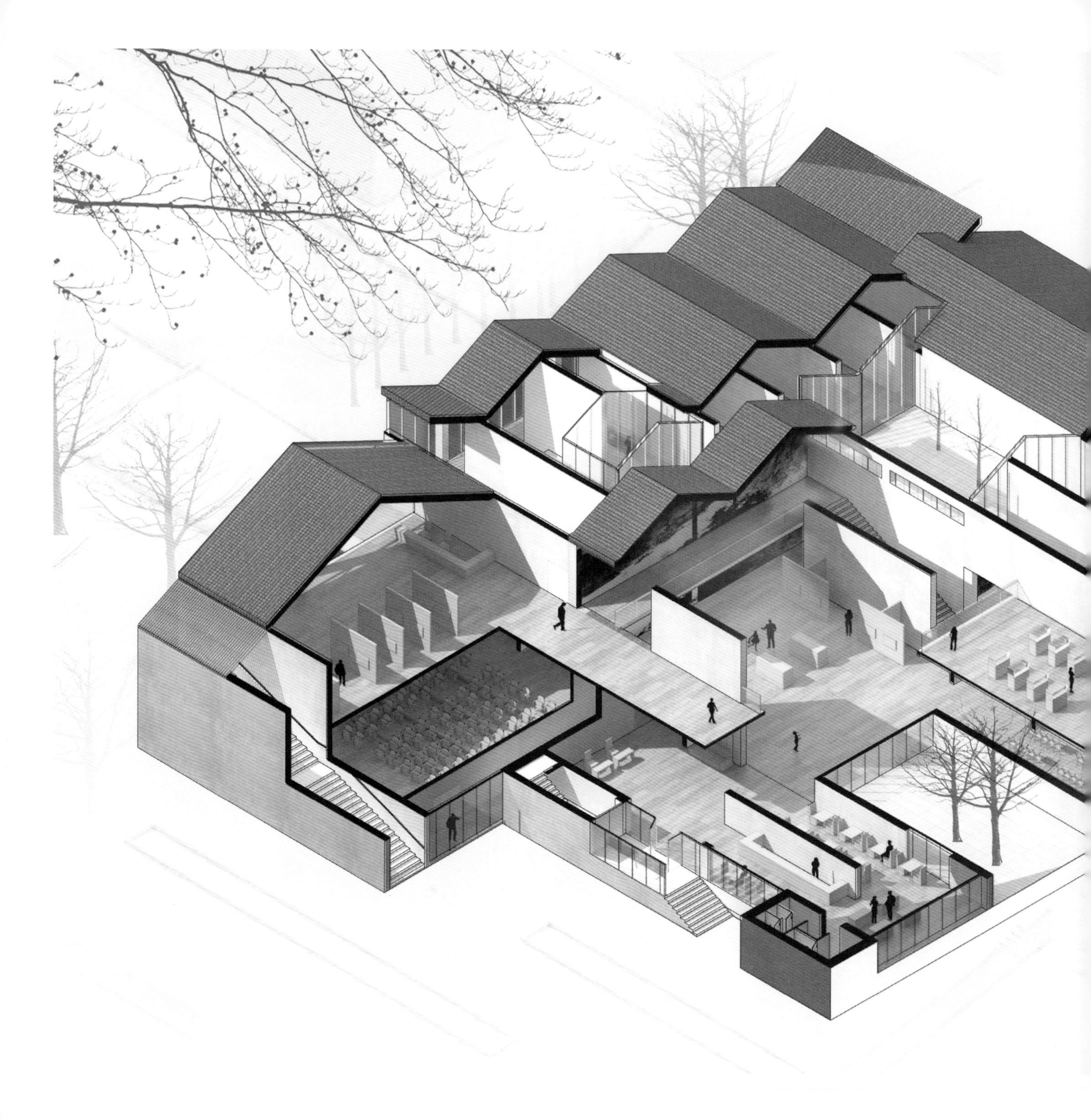

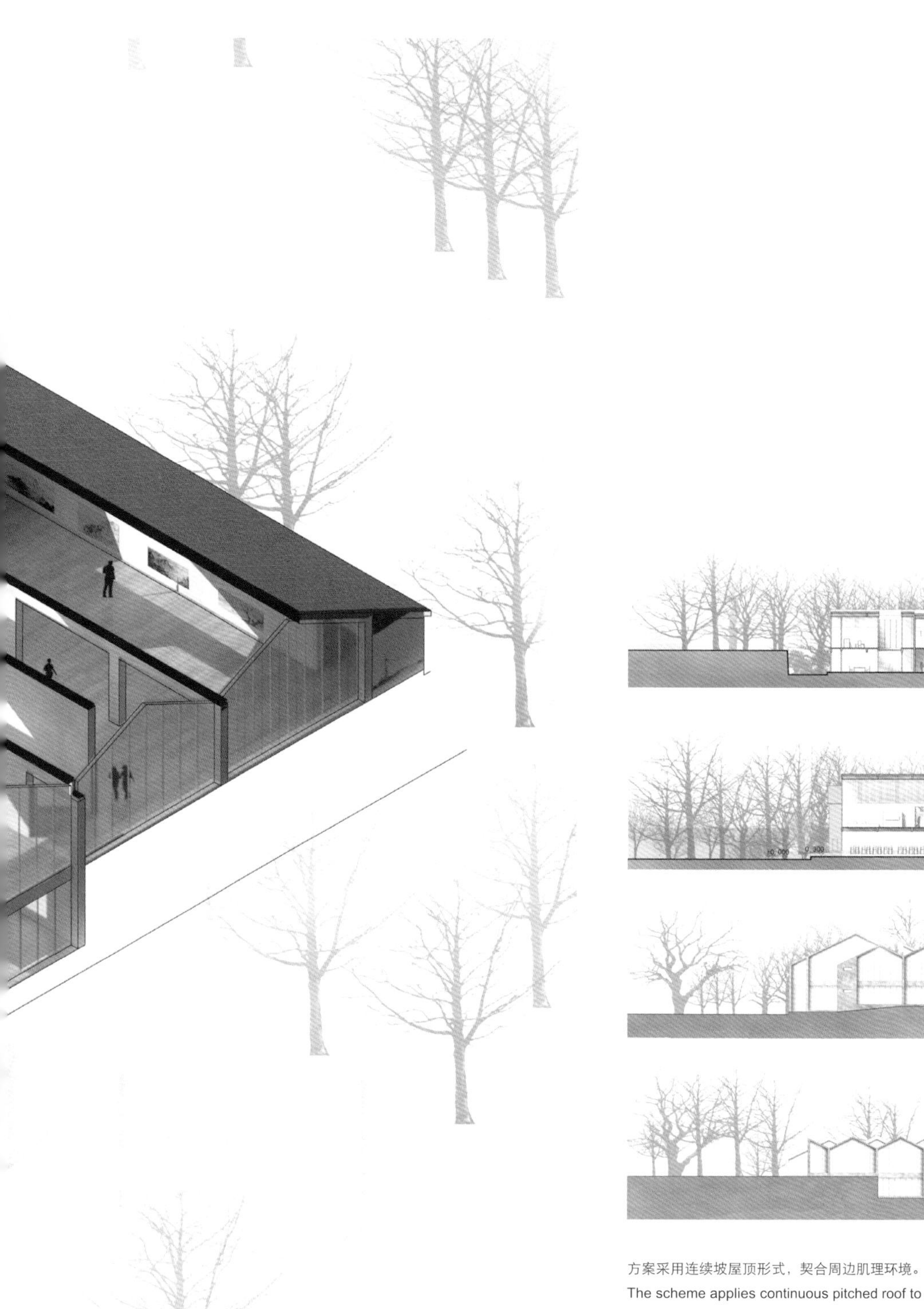

方案采用连续坡屋顶形式，契合周边肌理环境。

The scheme applies continuous pitched roof to match the surrounding texture and environment.

剑 阁 路
藏品库
工作间
工作间
会议室
办公室
设备间
设备间
次入口
报告厅
准备间
展厅
门厅
±0.00
临时展厅
主入口
一层平面 1：200
次入口
咖啡厅
厨房
办公室
办公室
办公室
设备间
设备间
创作间
创作间
创作间
创作间
创作间
创作间
展厅
5.20
二层平面 1：200
汉 口 西 路

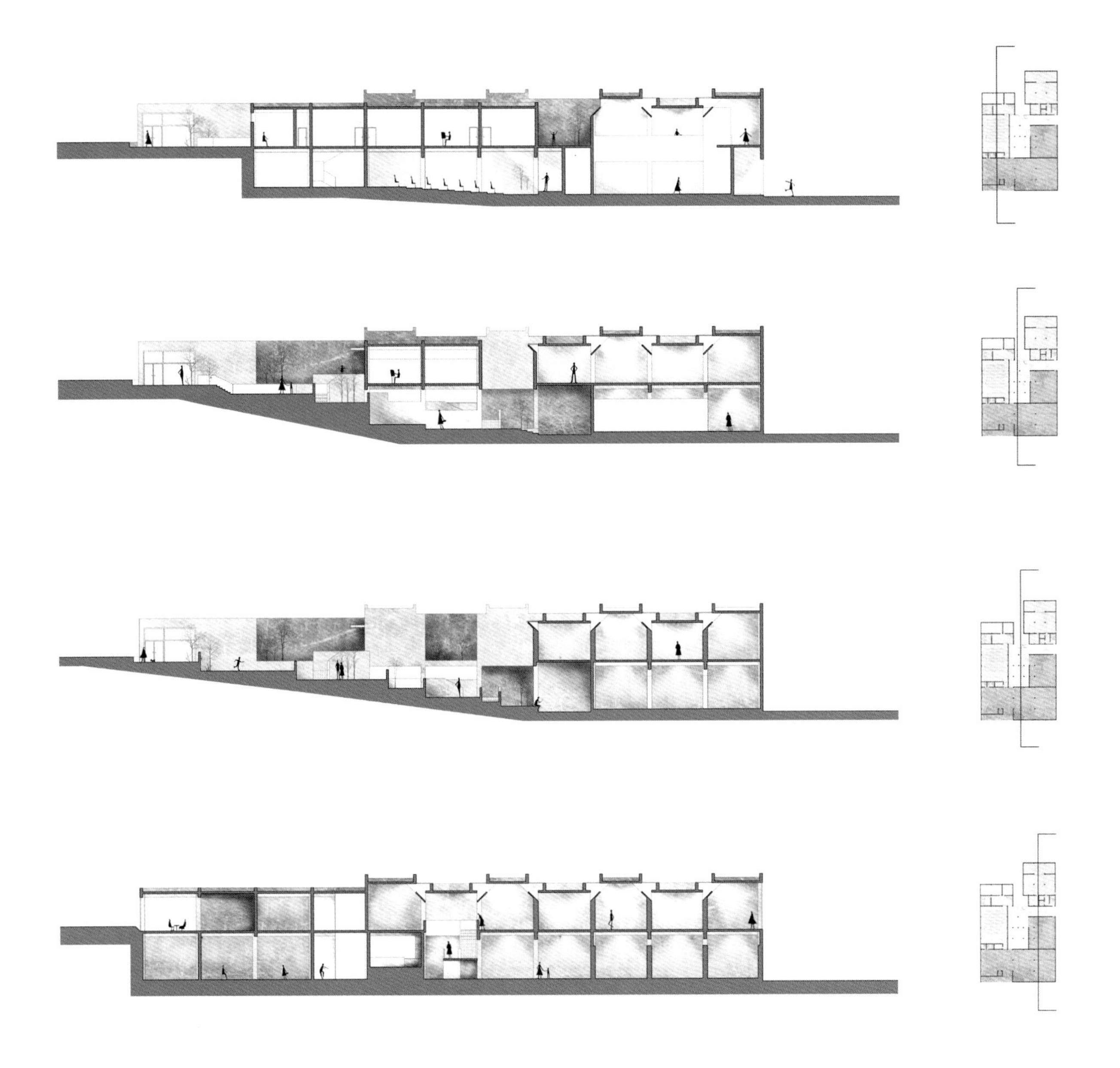

运用单元化的形体适应不同的功能需要，通过屋顶高度的变化及反光构件的设置为三个连续的展览单元采光，在垂直于单元展开的方向上营造不同的光照效果的序列变化。

The design applies modular form and structure to meet different functional requirements, and provides lighting for three continuous exhibition units via changing the roof height and setting the reflective components, creating sequence changes of different lighting effects along the direction perpendicular to the stretching direction of the units.

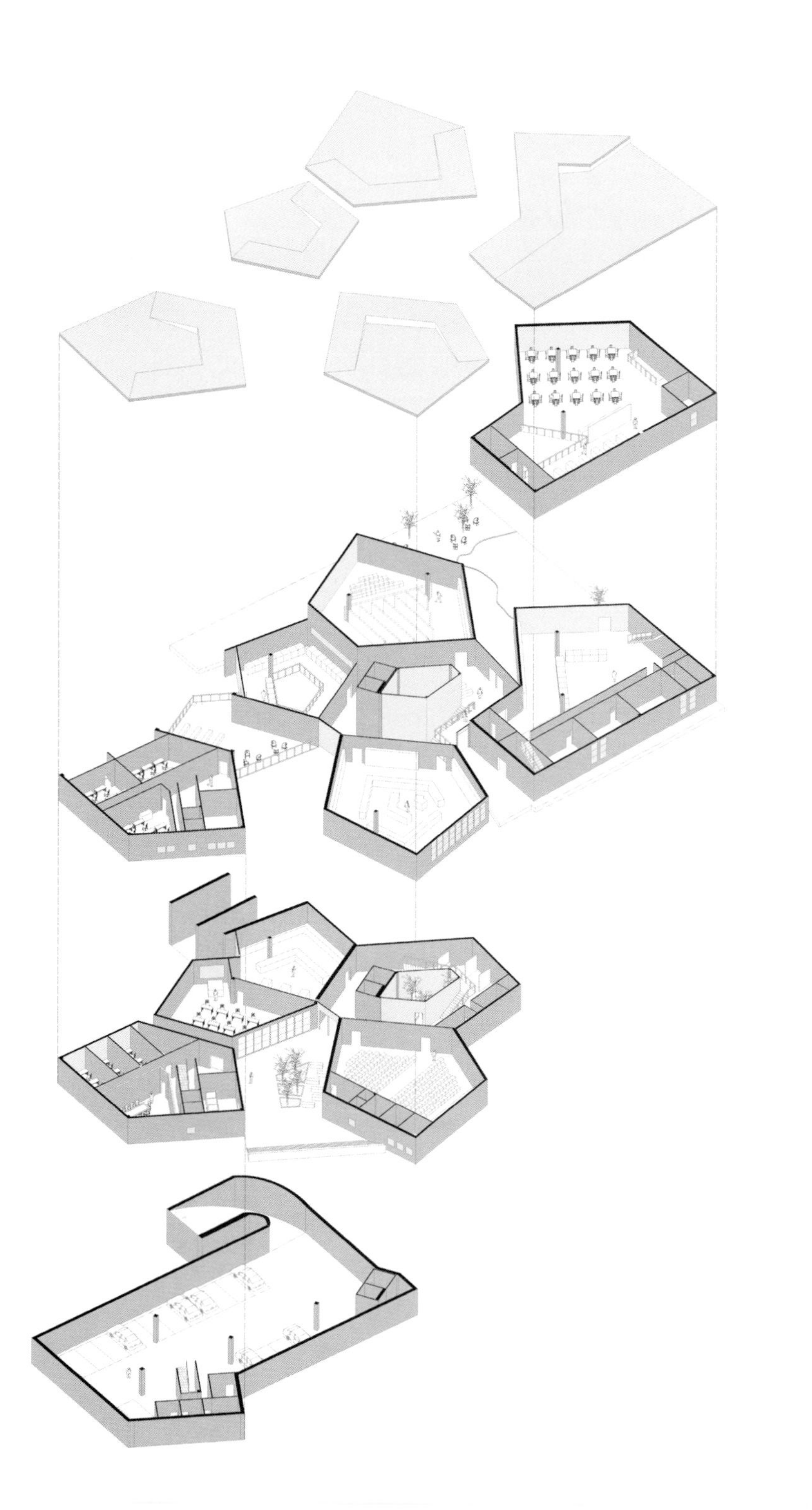

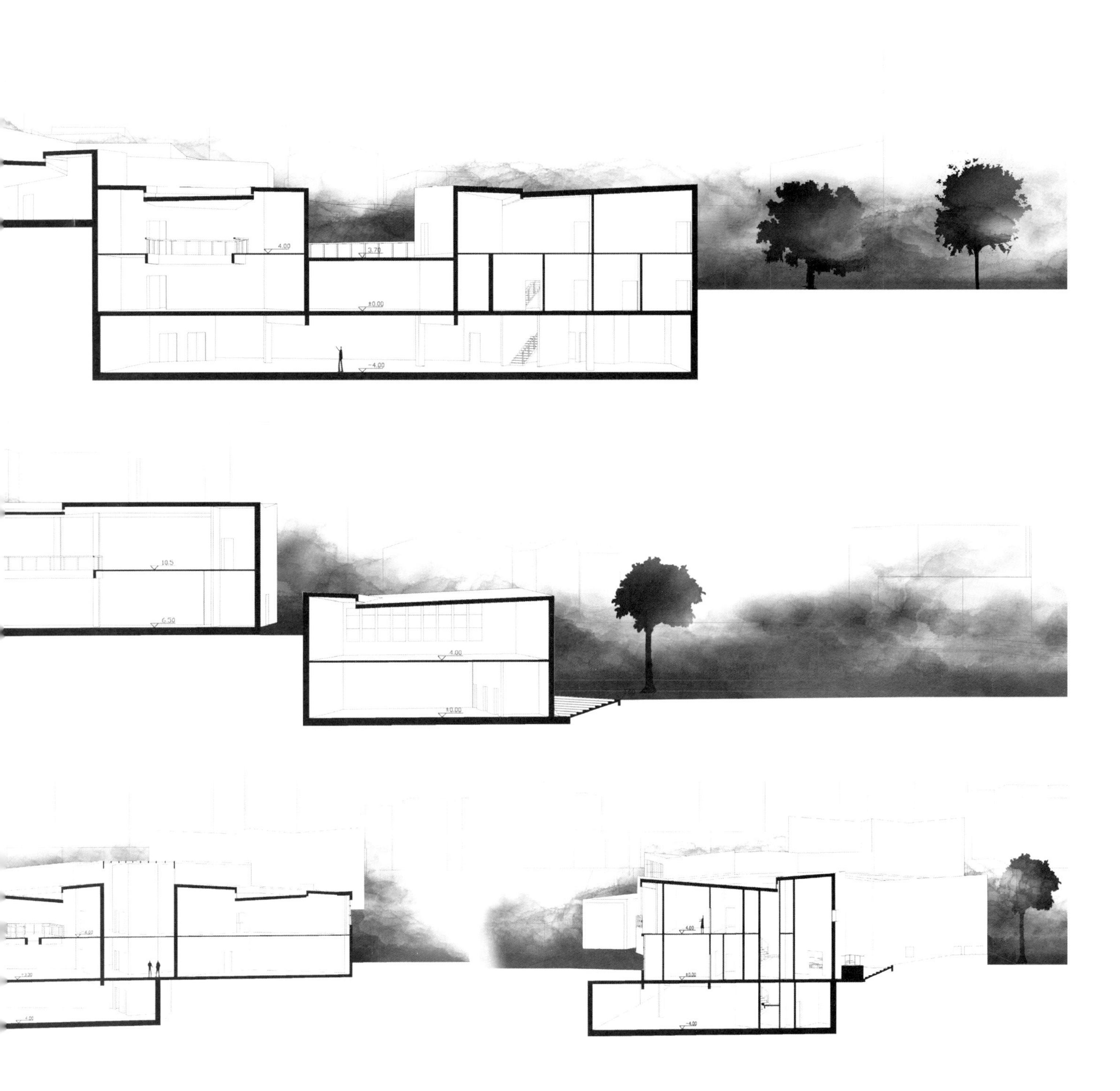
4.00
3.70
±0.00
-4.00
10.5
6.50
4.00
±0.00
4.00
±0.00
-4.00

城市建筑：社区中心

URBAN ARCHITECTURE: COMMUNITY CENTER

胡友培 钟华颖 王铠

城市已经成为当代建筑学无可逃避的语境，建筑学实践也成为城市生活不可或缺的一部分。对于处于三年级春季学期的学生，尽管经过了两年专业教育，但对于建筑学的观念，大多还局限于单体建筑。受到较为狭窄的生活经验所限，所谓都市生活、都市建筑问题对他们都还显得抽象而难以理解。

为处于该阶段的学生引入城市建筑的相关概念，实现认知与技能的提升、进阶，是本课程的根本任务。具体的，在观念上引导学生建立起城市区位与场地、街区与地块、街道界面与建筑立面、规划设计要点与指标、城市交通与场地微交通、公共建筑的尺度与公共性等概念。在技能上，则训练大中型公共建筑的设计能力——复杂功能的组织、内部流线与场地流线的规划、公共建筑立面语言、各种空间尺度与比例的掌握等。

如何将上述教学目标落实为具体操作环节？如何弥合三年级同学所处的知识水平与课题设定目标之间较大的落差？需要找到一个恰当的选题——社区中心，以及更为根本的，一种设计操作的方法——空间的方法。

社区中心，天然地具有了一个城市建筑所要训练的若干方面：一定的规模、足够的内在功能复杂性、超出建筑单体范围的城市职能、建筑的公共性、场地的多向异质性等，是一个较为理想的训练载体。作为三年级下学期的设计题目，该题材已经历时数年。本次课程沿用该题目，在具体的功能组成、面积规模上略有调整。更为重要的改变与尝试，则出现在设计操作方法上。

空间的方法，是本次课程为学生准备的主要设计工具。空间议题，似乎是现代主义的老生常谈，与当代都市建筑学话语略显违和。然而，我们认为无论当代都市情况的瞬息变化而导致的建筑学外部性领域(exteriority)如何动态与激变，建筑学的内部性(interiority)始终具有相对稳定的内容。空间的议题，则是其中之一。这既是现代主义的遗产，也是我们今天展开所谓当代建筑学话语的一个基础。课程制定了一条以空间为主线的、逐渐推进的四阶段课程计划，依次为：场地-空间、功能-空间、结构-空间、材料-空间。

场地-空间：对应常规设计流程的场地规划阶段。在此阶段，除了需要解决基本的场地规划问题，如动线组织、功能区划、出入口设置等，更需要关注场地的空间维度。该阶段引导学生认知、判断自己设计的建筑形体，在场地以及与周边建筑体量相互影响下，限定出的多种城市场所空间；引导学生思考这些场所可能承载的社会活动与身体行为，从而在身体尺度和个体经验上理解自己的设计。比例1：1000，体量模型＋场地模型。

功能-空间：对应一般设计流程的“排平面”阶段。由现代主义发展而来的各种空间方法，一个重要的贡献在于使得设计工作可以突破功能房间的束缚而开展空间的想象。从这点出发，强化平面阶段的空间内容，将有效摆脱各种类型化设计、功能主义所产生的机械性和约束性。基于现当代建筑学的丰厚遗产，课程按照一定的历史顺序，依次向学生引介了柯布西耶作品中的“空间的透明性”、康的“空间的秩序”、库哈斯的“空的策略”以及王澍的“作为肌理的建筑”。当然，这些具体的空间技术与方法，并未穷举所有重要的空间方法。然而，就课程训练而言，四种方法为学生提供了明确的空间话语方式，使其原本下意识的、混沌的空间形式操作得以明晰而有所理据，提升了学生的空间能力——想象与理解。比例1：500~1：200，空间结构模型（图）。

结构-空间：结构在这里指建筑结构形式。课程引导学生在空间的层面思考结构问题。由于在功能任务里有意安排了一个大跨度空间（游泳馆），使得结构问题成为本次设计无法回避的一个工作内容。通过案例的分析以及讲座，使得学生理解结构所产生的建构美学及其空间效果，并最终在自己设计中予以尝试。比例1：50，室内结构模型。

材料-空间：对应方案深化、详图阶段。由于是16周的长设计，使得课程有了一般8周设计所难以获得的深入设计的机会。因此，在材料与详图方面，课程提出了更高的要求，也使得学生在本科中期阶段，有机会较为全面地接触设计实践的全部流程。材料的引入，并不是单纯地上构造课（三年级有相关课程进行专门知识讲解），而是再次以空间为主线展开，重点围绕材料带来的空间氛围进行讲解。在个人设计工作中，则要求学生绘制大比例立面详图、重点室内空间渲染，以此为媒介推进方案的深化设计。比例1:20~1:50，渲染图或室内模型。

作为课程的总结，可以看到，以空间为主线、引入城市建筑的相关观念与技能具有成效。这表现在：同学开始使用具体的空间话语，展开城市环境与建筑单体相互关系的讨论与陈述。这避免了一些在讨论城市时常常使用的大词的陷阱（如城市结构、轴线、系统等），而回到了具体的身体、经验层面，如一片用绿地软化的广场、一个具有开放姿态的街道界面、一个建筑体量的转折，以及如何融入并呈现已有的城市环境。另一方面是不足与遗憾。这表现在“由外而内”与“由内而外”上。由于场所-空间、功能-空间的阶段性，使得部分同学并不能很好地实现内外建筑与城市空间的转化与互动。在第二阶段内部空间秩序的塑造时，往往自顾不暇，而难以再回到外部空间视角审视自己的设计。相反，在后两个阶段中，由于主题间先天的紧密联系，使得同学在局部深化推进时，仍能保持一个以单体内部空间为视域的平衡感与整体感。这里隐含的一个深层问题是：在城市建筑的训练中，空间作为主线的合法性。空间适合作为城市建筑的训练的工具吗？如何在城市建筑的层面，拓展现代主义基于内视发展而来的“建筑空间”概念与方法？如何进一步从当代建筑学中总结并提取出“城市建筑空间”的观念与方法？这些问题将保持开放，并可成为我们课程设计持续探讨的议题。

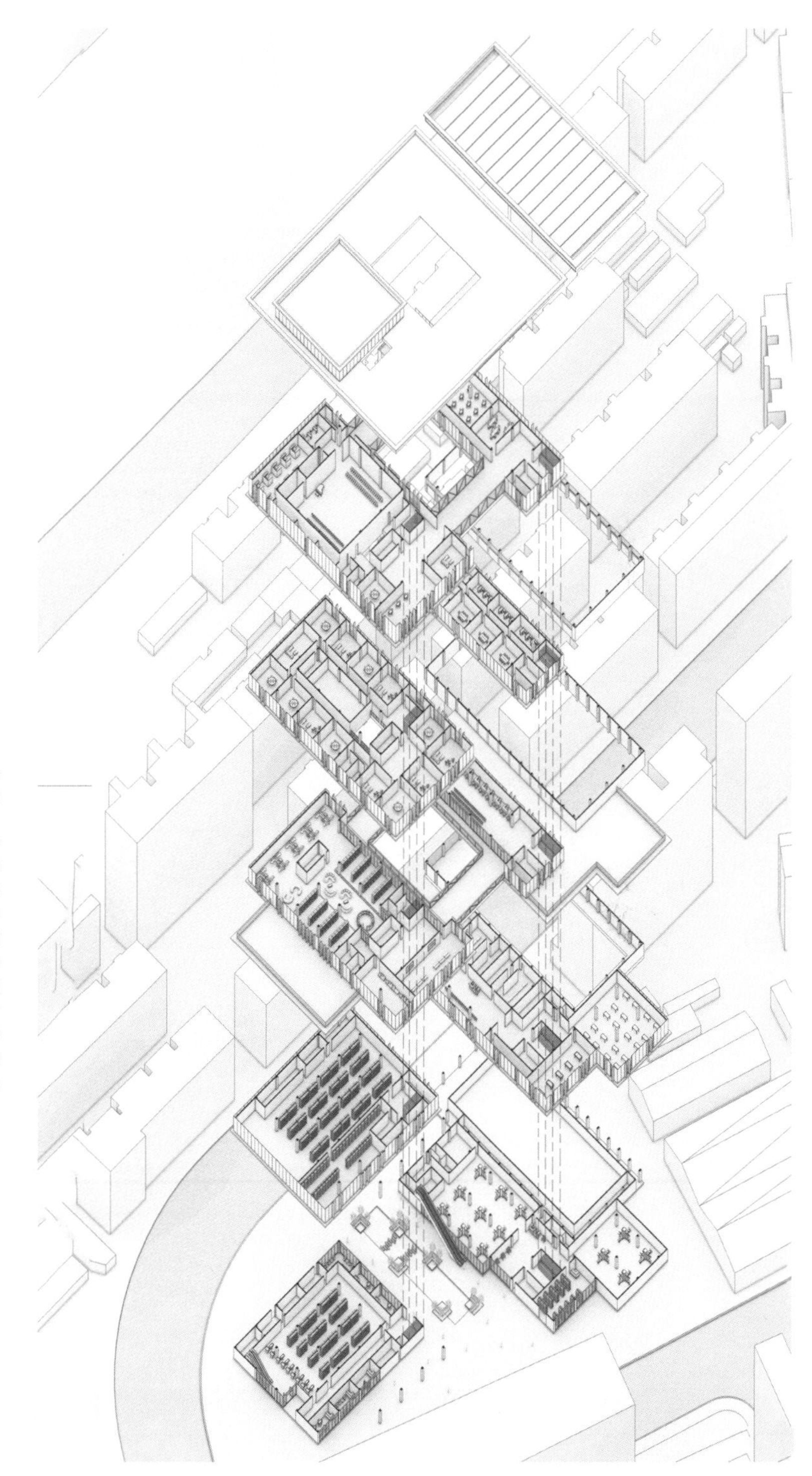

City has become an inevitable context for modern architecture, and practice of architecture has also become an integral part of urban life. For undergraduates in spring semester of their third year, although they have received two years of professional education, their idea of architecture is still limited mostly to individual buildings. Due to the limit of relatively confined life experience, the issues of so-called urban life and urban architecture are still abstract and difficult for them to understand.

For students in this phase, introduction of concepts related with urban buildings to realize improvement and upgrading of perception and skills is a fundamental task for the course. Specifically, it guides students to establish concepts such as urban location and field, block and land parcel, street interface and building facade, key points and indicators for planning and design, urban traffic and field micro-traffic, dimensions and public nature of public buildings in term of perception. And in term of skills, it trains them the ability for design of large and medium-sized public buildings – organization of complex functions, planning of internal circulation and field circulation, language of facade of public buildings, and grasp of dimensions and ratios of various spaces.

How could we turn above teaching objectives into specific operating links, and how could we bridge the gap between knowledge level of third-year students and objectives set out in this course, it requires us to select an appropriate topic – community center; and more fundamentally, one way of design operation – method for space.

Community center has the nature in multiple aspects that are required for training of an urban building: certain sizes, sufficient complexity of internal functions, urban function beyond the scope of individual building, public nature of the building, and multi-directional heterogeneity of the field, and it is an ideal training carrier. As a design subject for second semester of the third year, this topic has been lasted for years. The subject is continued in this course, with some slight adjustments on composition of specific functions, area, and sizes. And some more important changes and trials are presented on design operation method.

Method for space is the main design tool prepared for students by this course. The topic of space seems like a commonplace talk of modernism and in certain degree is disharmonious with the discourse of modern urban architecture. However, in our opinion, no matter what dynamic and violent changes of exteriority of architecture are caused by instantaneous change of modern urban conditions, interiority of architecture always maintains relatively stable content. The topic of space is one of them. This is not only the legacy of modernism, but also the foundation of the so-called the modern architectural discourse carried out today by us. A four-phase course plan that makes space as the main line and advances gradually is established for the course. They are in sequence: field-space, function-space, structure-space, and material-space.

Field-space: It corresponds to the field planning phase of conventional design process. During this phase, in addition to resolutions for issues of basic field planning, such as organization of circulation, functional zoning, arrangement of access and exit, more attentions are paid to dimensions of space. Guide students to recognize and determine spaces at various urban places defined by the architectural forms design by themselves due to effect between field and mass of surrounding buildings. Guide students to deliberate social activities and physical behaviors to be borne by such places, and to understand their design on basis of physical dimensions and personal experience. Scale 1∶1000, massing model + field model.

Function-space: It corresponds to the "plan design" phase of general design process. One of important contributions of various methods for space developed from modernism lies in that it enables design work to break through constraint of functional rooms, and unfold imagination on space. Starting from this point, and emphasizing content of space during plan design phase, it can effectively get rid of mechanism and constraint generated by various typified designs and functionalism. On basis of the abundant heritage of modern and contemporary architecture, the course introduces students, in certain historical sequence, with concepts of "transparency of space" in works of Corbusier, "sequence of space" by Kahn, "strategy of space" by Koolhaas, and "building as texture" by Wang Shu. Of course, these specific space technologies and methods do not cover all important methods for space. However, for course training, these four methods provide students with specific modes of spatial discourse, make their subconscious, chaotic space form operations become clear and rational, and improve space ability of students – imagination and understanding. Scale 1∶500~1∶200, model of space structure (drawing).

Structure-space: Here the structure refers to structural form of buildings. The course is intended to guide student to deliberate structure issue at space level. Since a large-span space (swimming pool) is arranged intentionally in function task, it makes the issue of structure become an inevitable work content in this design. By the way of case analysis and lecture, it allows students to understand tectonic aesthetics and its spatial effect generated by structure, and to try it in their own design in the end. Scale 1∶50, model of interior structure.

Material-space: It corresponds to deepening of plan, and detail drawing phase. Given that it is a long-term design process spanning 16 weeks, the course has the opportunity of making in-depth design that is generally unavailable for an 8-week course. Therefore, the course has higher requirements in terms of materials and detail drawings. It also offers students an opportunity to engage in the entire process in an all-round way in the middle phase of their undergraduate education. Introduction of materials is not just a course of pure tectonics (relevant knowledge has been taught specially at associated courses in the third year), but unfolding the main line of space again. It focuses on interpretation on space atmosphere generated by materials. In personal design, it requires students to complete large-scale detail drawings of elevations, focusing on rendering of interior space, and push forward in-depth design by making it as the medium. Scale1∶20~1∶50, renderings or interior models.

In conclusion of the course, we can see that by making space as the main line, introduction of ideas and skills about urban buildings is effective. The performance lies in: students began to unfold discussion and presentation on mutual relations between urban environment and individual buildings with specific space discourses. It avoids the use of large words (such as urban structure, axis, system, etc.) that are used frequently in discussion of a city, and go back to the level of specific human body and experience, for example, a square softened with green land, a street interface with gesture of openness, a transition of building mass, and the way to blend in and present existing urban environment. On the other hand, it also has some shortcomings and pity. This is displayed in terms of "from exterior to interior" and "from interior to exterior". Due to different phases of field-space and function-space, some students could not master the translation and interaction between interior and exterior of buildings and urban space. When interior space sequence is shaped in the second phase, students often are too busy to handle themselves, and feel difficult to go back and review their design from the angle of exterior space. On the contrary, in the last two phases, given the inherent close connection between topics, students could still maintain the sense of balance and entirety from the perspective of individual interior space when local detail design is push forward. It implies a deeper question: the validity of making space as main line in the training of urban buildings. Is it appropriate to use space as a training tool for urban buildings? How could we expand the concepts and methods of "building space" that are developed by modernism based on interior view on the level of urban buildings? How could we further conclude and extract ideas and methods about "urban building space" from contemporary architecture? These questions will still maintain open, and become topics of our future discussion on course design.

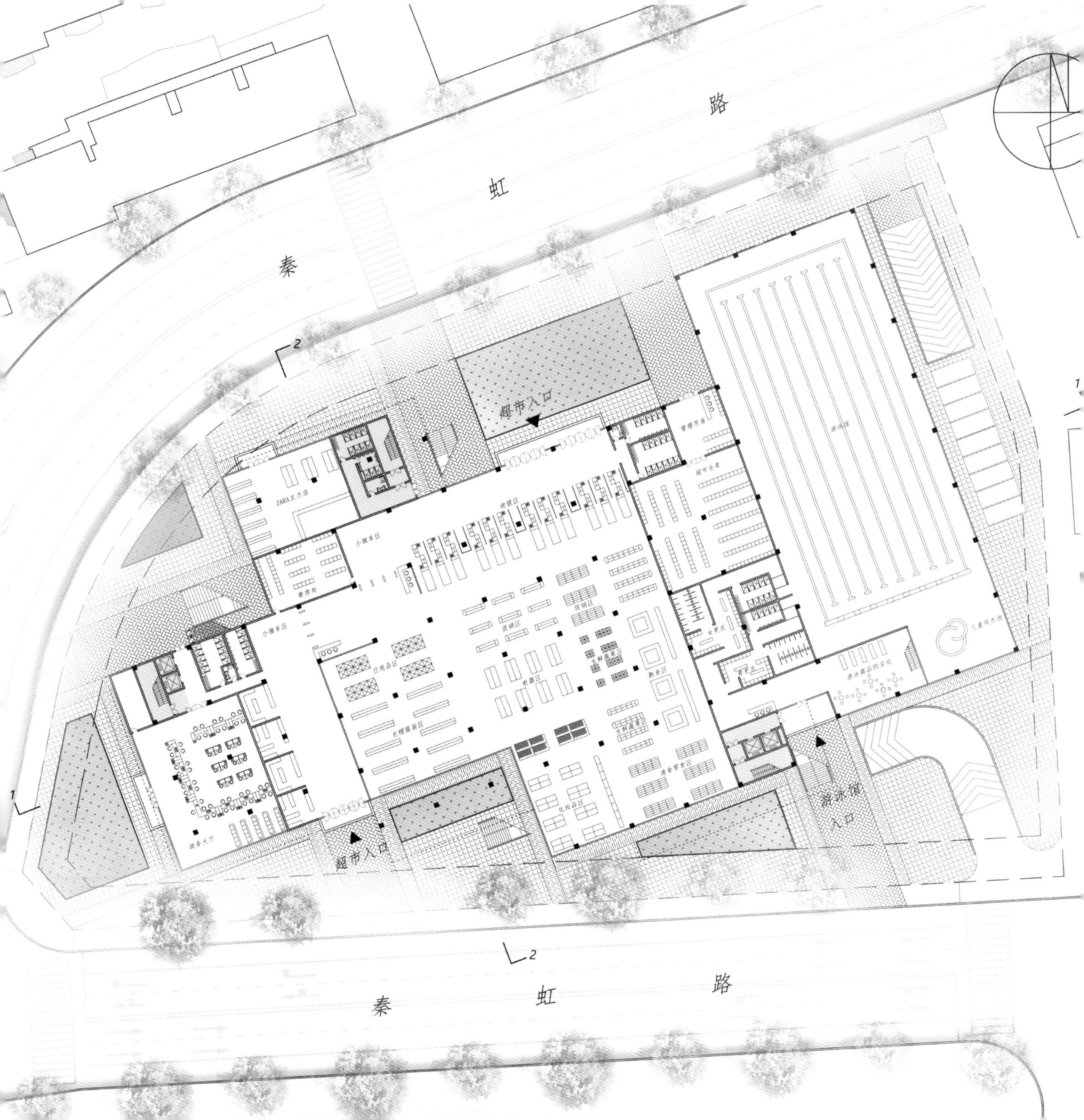

秦 虹 路
超市入口
ZARA主力店
管理用房
超市仓库
游泳馆
收银区
小推车区
寄存处
促销区
日用品区
电器区
生鲜蔬果区
熟食区
女更衣
男更衣
儿童戏水池
游泳商品购买处
衣帽服装区
生鲜蔬果区
酒食零食区
化妆品区
政务大厅
超市入口
游泳馆入口
秦 虹 路

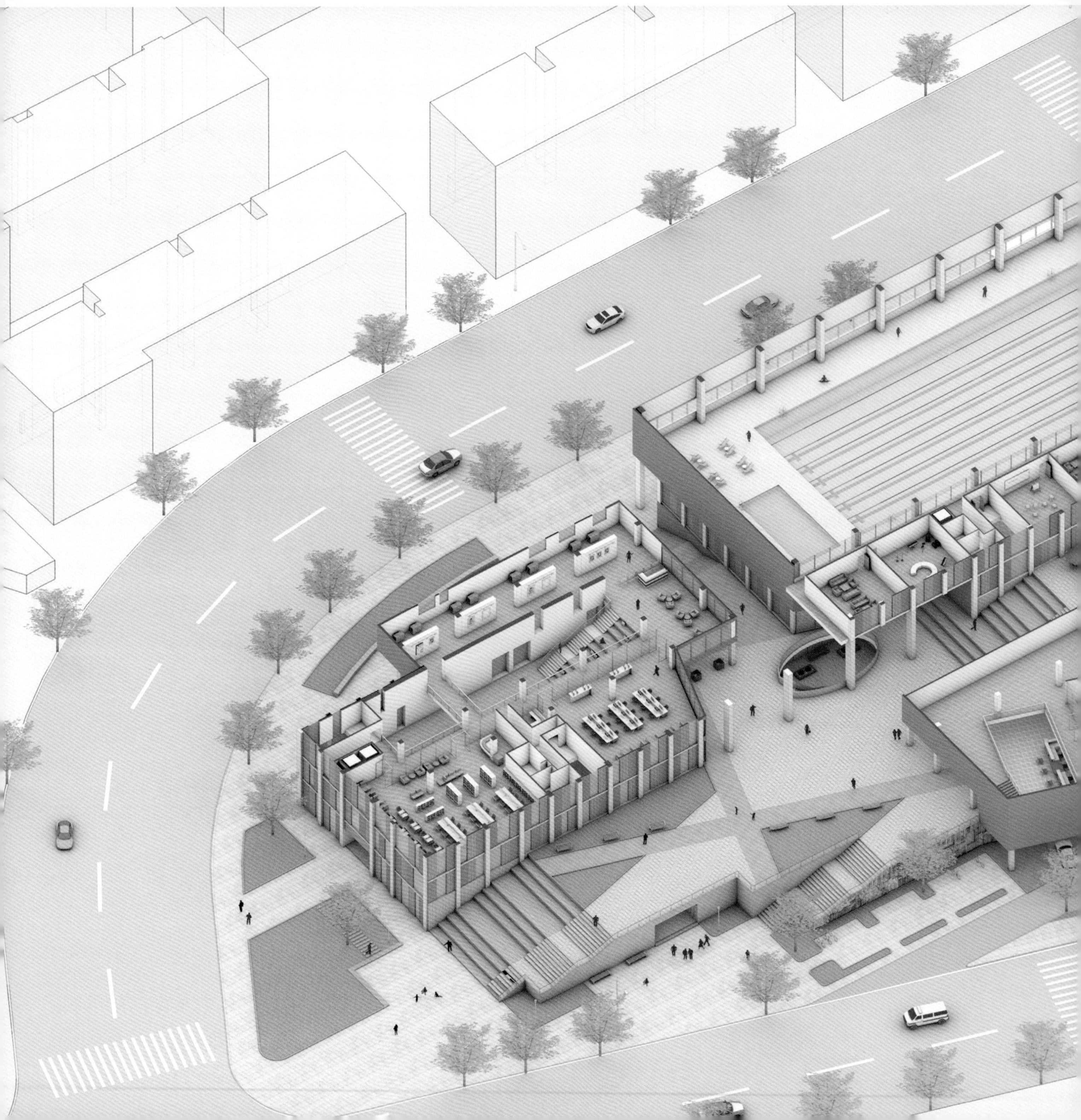

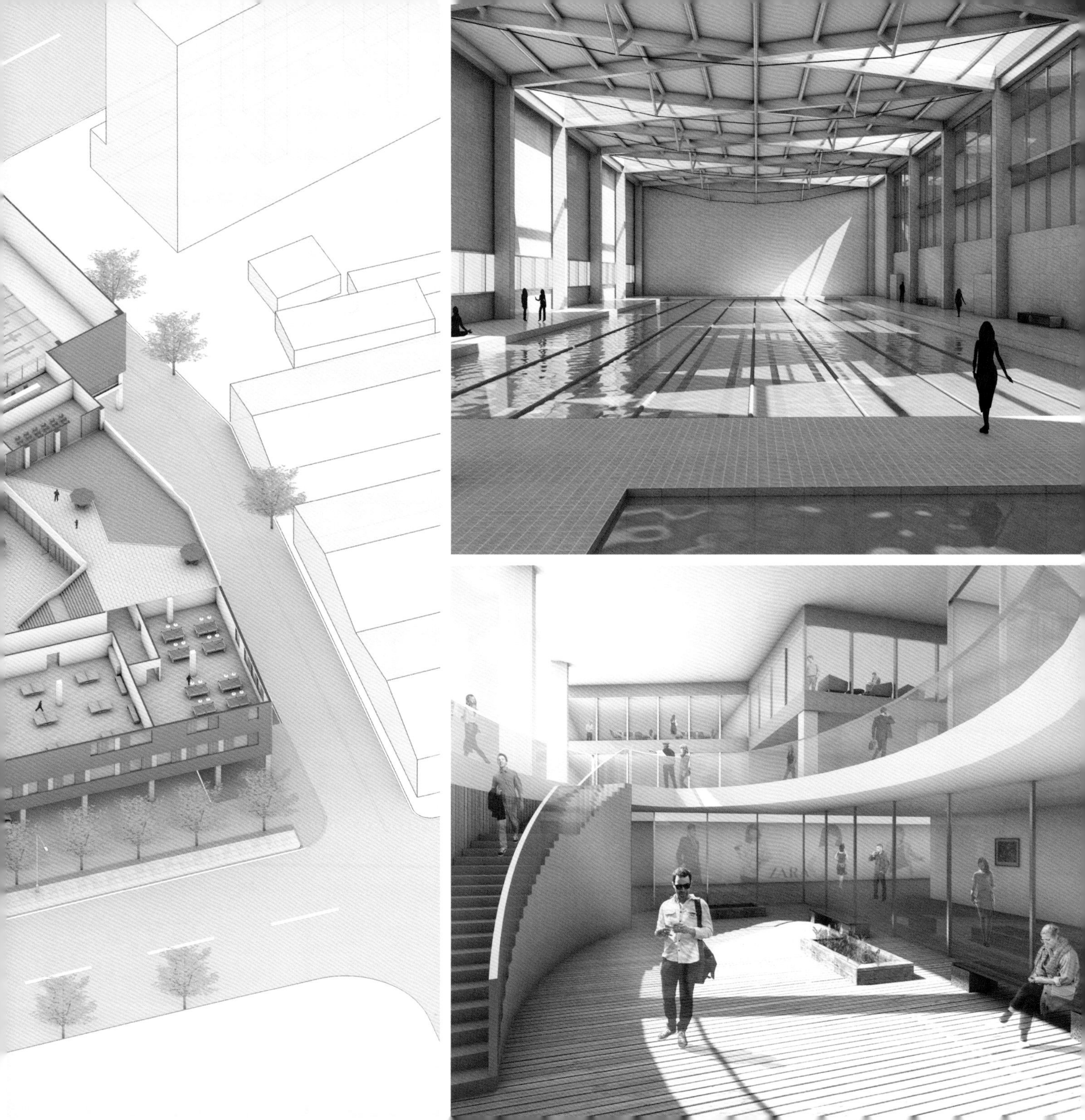

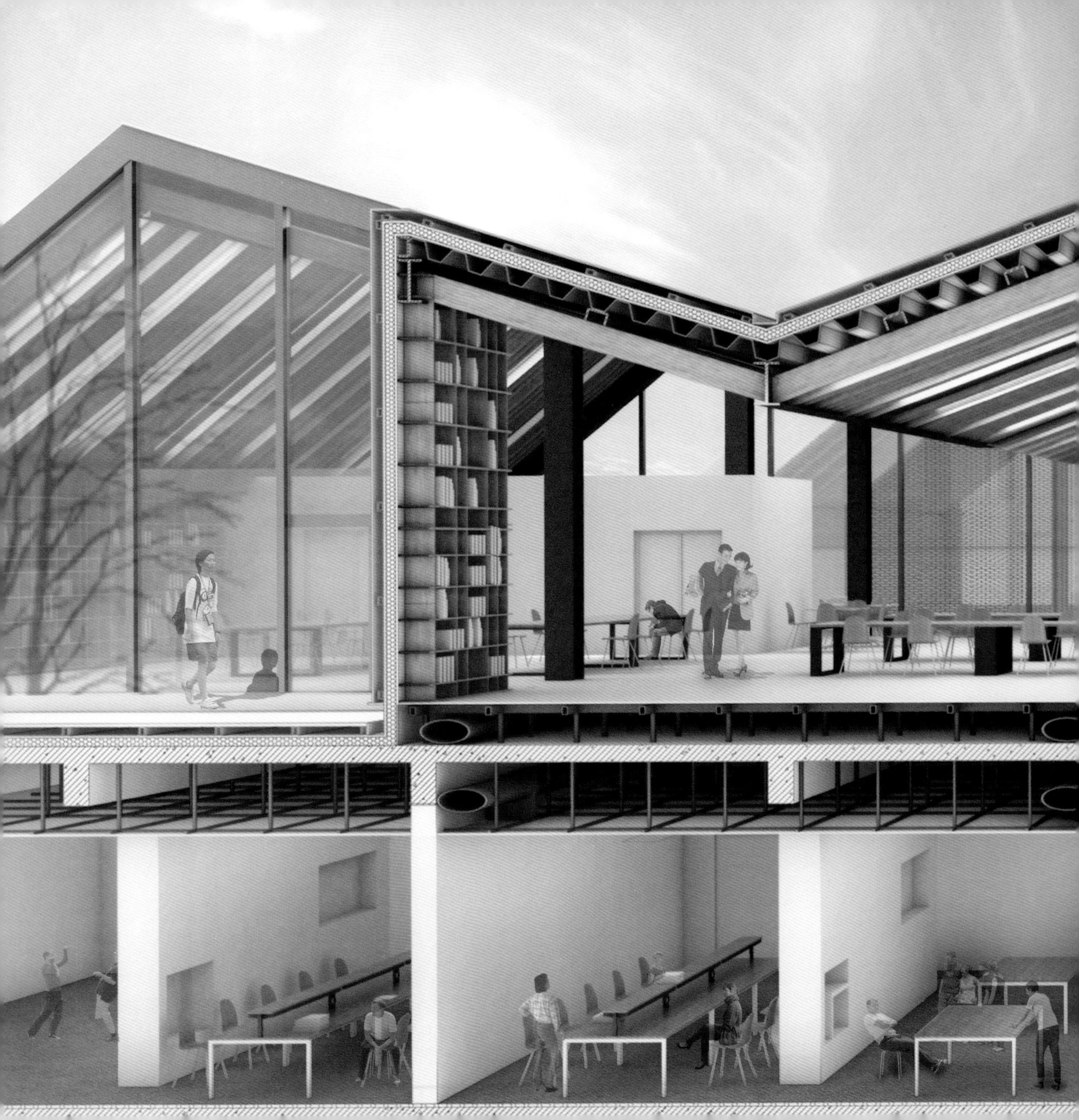

旧城改造城市设计
URBAN DESIGN FOR OLD TOWN RENOVATION

吉国华 胡友培 尹航

城市建筑不应该只是以自我为中心的独立个体，它们需要融入周边的整体城市空间环境。认知建筑新的角色、理解建筑在城市环境中的位置，是建筑师必备的技能。建筑学本科的最后一个课程设计——“城市设计”就是为建筑学本科同学开启通向这全新知识领域的道路。我们在城市设计课程中选择的环境是学生熟悉的城市中心区，训练学生对城市环境的调研方法，引导学生超越之前建筑单体设计中对内部空间与形态的执着，并逐步从城市整体容量功能与外部空间的角度去整合建筑。

1. 训练目标

我们为“城市设计”课程设置的训练主要目标有：

（1）着重训练城市空间场所的创造能力，通过体验认知城市公共开放空间与城市日常生活场所的关联，运用景观环境的策略创造富于空间特征的城市空间；

（2）建立起城市基础设施的基本概念，通过对设计场地交通、水电、竖向排水等设施的现状调研与更新设计，了解城市层面的基础设施网络体系与建筑的关系；

（3）掌握城市设计的方法，初步形成从宏观整体层面处理不同尺度空间的能力，并能有效地用各类型图纸进行表达；

（4）了解城市更新的基本概念，结合城市中心区保护建筑与保留建筑的不同处理方式，理解复杂城市环境中城市更新的机制和价值；

（5）培养团队合作意识和分工协作的工作方式。

2. 设计课程分阶段

“城市设计”是本科同学们第一次面对如此真实与复杂的场地和一种全新的设计对象，因此我们设置了严格的设计分期，让学生尽早地完成从建筑设计到城市设计的思维转换。课程大致分为以下六个阶段，用八周的时间完成。

2.1 场地分析

（第一周——调研测绘—上位规划解读—环境要素提取—相关案例调研）

我们为城市设计选择场地的原则是：公共功能主导、适度的功能复杂性、调研的便捷度等。2015年课程中我们选取的仍然是太平北路—珠江路—碑亭巷—长江路这四条路围合的地块，为了增加地块展开深度，增加美术馆和南京烟厂，共约6.2公顷（不含现状城市道路）。学生需用一周时间完成初步调研与分析。

学生需要调研的城市环境特征大致体现在以下几个方面：区位环境、用地性质、道路交通、原建筑质量、文化资源、相邻地块情况、现状停车与地下空间系统等。

根据之前的教学经验，我们还要求学生对场地居民和未来可能的使用者以及周边相邻居民进行问卷调查，完整地梳理出地块现状的问题与机遇。第二周第一节课进行分组调研汇报。

2.2 概念设计

（第二周——分析城市设计影响因素—发展城市设计策略）

（第三周——塑造城市公共空间体系）

在初步调研分析之后，学生用两周的时间进行概念设计，各组在自己分析出的场地缺陷与优势中找出主要的城市设计影响因素，组内成员先独自提交若干条可能的城市设计发展策略，在第二周进行初步的概念总平面设计，经过组内讨论和老师评图，进行组内筛选和整合，并在第三周发展出具有完整公共空间体系（如步行系统、开外空间系统、绿地系统等）的初步方案，形成城市形态的结构性框架并初步调研分析选定符合环境特征的城市肌理，以体积模型与总平面图示的形式在第四周初进行班级讨论。

2.3 城市空间组织

（第四周——城市公共空间体系的细化）

学生用一周的时间，对概念设计确定的城市空间体系进行细化设计。本周需要确定公共空间体系的主要节点与轴线、相应的主次干道，并以此为原则合理划分地块；组织分地块间的交通联系与地块内部的人车流线；进一步深化城市肌理，优化建筑体量与总体形态控制。

本周各组需完成内部成员的概念整合，确定一定的分工，提交细化到单体建筑体量的体积模型（1：1000）、概念设计总图（1：2000）、反映街巷空间视线的总体剖面、能说明公共空间体系组织的分析图或序列透视图等。

2.4 建筑类型与肌理

（第五周——城市空间的建筑类型学翻译）

确定了公共空间体系后，学生需对第三周初步得出的城市肌理进行建筑类型学翻译，这周的工作需要运用在建筑单体设计训练中养成的对功能、面积和形态的敏感度以及整体把握能力，但是又要避免陷入对单体形态与细部的纠缠。

本周在完善城市空间体系的模型与各项图纸之外，需对建筑肌理控制提出具体的来源与分析过程，用概念模型或图纸表达建筑类型的生成分析。这周训练学生们将城市空间与建筑形态特征进行总体整合，引导他们从对城市周边现状肌理的分析和特定城市功能的建筑空间生成等方面着手，初步了解建筑类型学的机制，完成从建筑单体设计到整体城市形态控制的转换。

2.5 环境与公共空间节点

（第六周——场地环境与公共空间的深化设计）

本周的设计任务是对城市公共空间系统的分解与深化，对重要的空间节点进行深化和景观试做，这周的训练主要是对公共空间体系的合理性进行最后审视，通过示范性的景观深化设计，从视觉上进行展示与验证。

本周还需要对城市基础设施进行系统设计，解决城市设计的底层条件问题。

2.6 设计深化与表达

（第七、八周——设计深化与图纸定稿）

城市设计是建筑学本科最后一个课程设计，且由每组4~5人合作完成，因此最终的课程设计成果必须在数量和质量上达到一个新的高度，课程预留了两周的时间，给学生们将之前完成的城市公共空间体系、建筑类型、环境节点、基础设施等设计进行固化与表达。最终每组需要完成8~10张A0的图纸成果和相应的城市模型（1∶500）。

图纸包括但不仅限于：

（1）前期分析：上位规划解读、周边用地与交通、周边城市肌理、其他问卷调研等；

（2）基本图纸：城市设计总平面、用地平衡表、用地性质、总体鸟瞰和透视等；

（3）空间规划分析：区位分析、规划结构、道路交通、基础设施分布、更新策略等；

（4）空间设计分析：街道剖面、沿街立面、开放空间立面、场地剖面、重要节点表达等；

（5）分地块城市设计导则：地块描述、空间体量、街道界面（建筑退让街道缓冲空间的处理）、地块要求（退让、交通组织、停车）、建筑引导（建筑出挑、形式、材料、色彩、屋顶）。

3. 小结

对于逐渐着迷于建筑的个性化、创造性特征的本科四年级学生来说，城市设计课程是一针必要的清醒剂。学生们在这八周的课程中，会初步了解城市建筑的社会属性与城市属性。在复杂城市环境的调研、分析与更新设计中，他们会学习如何将建筑师对建筑空间与形态的创造欲望，与城市空间的公共性、既有城市环境的复杂性甚至城市更新的经济性等种种社会与城市要素进行结合。

在城市化进程不断加剧的当代中国，更凸显其必要性。

Urban buildings should not just be separate and self-centered individual ones, and they should be blended into entire surrounding urban space and environment. Recognizing new roles of buildings and understanding positions of buildings in urban environment are essential knowledge for an architect. The last course design for undergraduates of architecture – “Urban Design” is just the path leading undergraduate students of architecture to the brand-new field of knowledge. In the course of urban design, we select the environment of downtown area that is familiar to students, aiming to train students with investigation method on urban environment, and guide students to surpass the persistence on interior space and forms in the design of individual buildings, and to integrate buildings gradually from the perspective of overall capacity, functions, and external space of the city.

1. **Training Objectives**

We set up the following objectives for the training of the “urban design” course:

(1) Emphasize the training on ability of creating urban spatial places, and create urban space full of spatial features with the strategy of landscape environment through experiencing and perceiving the links between urban public spaces and urban daily living places.
(2) Establish basic concepts of urban infrastructure, and understand relations between infrastructure network and buildings from the angle of a city through investigation of existing conditions and renewal design of facilities such as field traffic, water supply, electricity, and vertical drainage.
(3) Master methodology of urban design, grasp the ability of handling spaces of different dimensions at macro and integral level, and achieve effective representation with drawings.
(4) Understand basic concepts of urban renewal, as well as the mechanism and value of urban renewal in complicated urban environment by combining different treatment ways of building protection and building preservation in downtown area.
(5) Cultivate awareness of teamwork and the working mode of collaboration.

2. **Stages of the Design Course**

"Urban design" is a brand-new design object being faced by undergraduate students for the first time that the site is so real and complicated, so we divide it into several stages strictly, so that students may successfully transfer their thinking mode from building design to urban design as early as possible. The course consists of the following six stages, which will be completed in eight weeks.

2.1 Sit Analysis

(Week 1: Investigation and mapping – reading superior planning – extracting environmental elements – studying associated cases)

Our principle for selection of the site for urban design is being guided with public functions, with appropriate functional complexity, and being convenient for investigation. For the course in 2015, we still selected the land parcel enclosed by North Taiping Road, Zhujiang Road, Beiting Road, and Changjiang Road, and in order to increase depth about the land parcel, we added the art gallery and Nanjing Tobacco Factory, covering an area of 6.2 hectares in total (excluding existing urban roads), and students were required to complete preliminary investigation and analysis within one week.

Urban and environmental features in investigation by students are mainly reflected in the following aspect: location and environment, nature of the land, roads and traffic, quality of existing buildings, cultural resource, conditions of adjacent land parcels, existing parking conditions, and underground space and systems.

According to previous teaching experience, students were also required to carry out questionnaire survey against residents at the site, potential future users, and surrounding residents, fully figure out existing problems and opportunities about the land parcel, and present their investigation reports in groups on first day of class in week 2.

2.2 Conceptual Design

(Week 2 – analyzing factors affecting urban design – developing strategy of urban design)

(Week 3 – Creating system of urban public space)

After preliminary investigation and analysis, students spent two weeks to complete conceptual design, each group figured out factors affecting urban design from pros and cons of the site according to their analysis, group members first presented their respective possible development strategies of urban design independently, and completed their preliminary conceptual master plans in Week 2, which were screened and integrated within the group after discussion by the group and being reviewed by the teacher. And in Week 3, they developed a preliminary plan with a complete system of public space (e.g. pedestrian system, open space system, and landscaping system), shaped a structural frame of urban form, and preliminarily investigated, analyzed, and determined the urban texture in accord with environmental features, which were discussed in the class in the form of volume model and master plan in early Week 4.

2.3 Organization of Urban Space

(Week 4 – Refining the system of urban public space)

Students spent one week to refine the design of the urban space system determined with conceptual design. In this week, students were required to determine main nodes and axis of the public space system, relevant primary and secondary roads, and to divide the land parcel reasonably on such basis; to organize traffic links between sub-land parcels as well as people and traffic flows within the land parcel; to further deepen urban texture, and optimize building volumes and overall form control.

During this week, each group was required to complete integration of concepts from its members, assign tasks, and submit volume model being refined to individual building volume (1:1000), master plan of conceptual design; overall profile that can reflect spatial view of lanes and streets, and analysis charts or sequence perspectives that can interpret the organization of public space system, etc.

2.4 Building Typology and Texture

(Week 5 – Translation of urban space with building typology)

After the public space system was determined, students were required to translate the preliminary urban texture achieved in Week 3 with architectural topology, and tasks in this week require the application of the sensitivity and overall manipulative ability on functions, area, and forms that are cultivated in design draining of individual buildings; however, they must avoid being caught in tangling with individual forms and details.

In addition to improvement of models and various drawings of the urban space system, students were required put forward specific source and analysis process for control of building texture, and express formation and analysis of building types with conceptual models or drawings. In this week, students were trained to carry out overall integration of urban space and building morphological characteristics, and were guided to start from aspects such as analysis on existing texture of surrounding urban area and formation of architectural space with specific urban functions, and to have a preliminary understanding of the mechanism of architectural typology, and complete the transformation from individual building design to overall control of urban form.

2.5 Details of Environment and Public Space

(Week 6 – Detail design of field environment and public space)

Design tasks in this week include breakdown and details of urban public space system, and complete in-depth design of key space details and trial modeling of landscape, and training in this week is focused on the final review on rationality of public space system, implementation of detail design with demonstrative landscape, as well as visual presentation and verification.

Students were also required to carry out systematic design for urban infrastructure, and solve problems of ground-floor conditions in urban design in this week.

2.6 Detailing and Presentation

(Week 7, 8 – Detail design and drawings finalization)

Urban Design is the last course design for undergraduate students of architecture, and it is completed jointly by a group consisting of 4~5 members, so the final design results of the course must reach a new level in terms of quantity and quality. The course has two weeks for students to finalize and present design works that have been completed in earlier stages, including the urban public space system, building type, environmental nodes, and infrastructure. And in the end, each group would complete design results of 8~10 drawings of A0 size and relevant city models (1:500). Those drawings include without limitation to:

(1) Early-stage analysis: Interpretation of superior planning, surrounding land parcels and traffic, surrounding urban texture, and other questionnaire surveys;

(2) Basic drawings: Master plan of urban design, land balance sheet, nature of the land, overall aerial view, and perspectives;

(3) Analysis on space planning: Location analysis, planning structure, roads and traffic, distribution of infrastructure, and renewal strategy;

(4) Analysis on space design: Street profiles, front elevations, elevations of open space, site profiles, and presentation of key details;

(5) Urban design guideline for sub-land parcels: land parcel description, spatial volume, street interface (treatment of street buffer space for building setback), requirements on land parcel (setback, traffic organization, parking), and building guidance (projection, forms, materials, color, and roof of buildings).

3. **Conclusion**

For senior undergraduate students who are gradually obsessed in personalized, creative features of buildings, the Urban Design course is a dose of sobriety. During this eight-week course, students preliminarily understood the social properties and urban properties of urban buildings, and through investigation, analysis, and renewal design about complex urban environment, learnt ways to integrate various social and urban elements with creative desire of architect on building space and forms, such as public nature of urban space, complexity of existing urban environment, and even the economic efficiency of urban renewal.

It further highlights the necessity in China that is increasingly urbanized today.

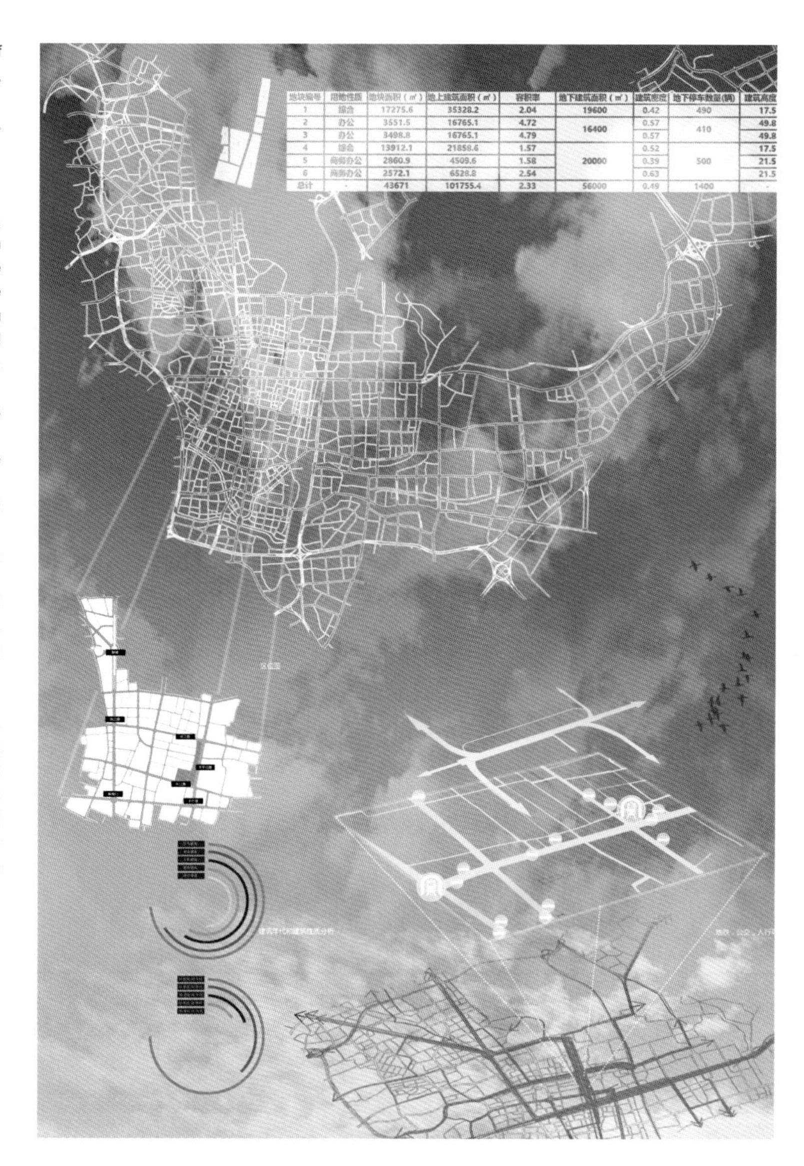

地块编号	用地性质	地块面积（㎡）	地上建筑面积（㎡）	容积率	地下建筑面积（㎡）	建筑密度	地下停车数量(辆)	建筑高度
1	综合	17275.6	35328.2	2.04	19600	0.42	490	17.5
2	办公	3551.5	16765.1	4.72	16400	0.57	410	49.8
3	办公	3498.8	16765.1	4.79		0.57		49.8
4	综合	13912.1	21858.6	1.57	20000	0.52	500	17.5
5	商务办公	2860.9	4509.6	1.58		0.39		21.5
6	商务办公	2572.1	6528.8	2.54		0.63		21.5
总计	-	43671	101755.4	2.33	56000	0.49	1400	-

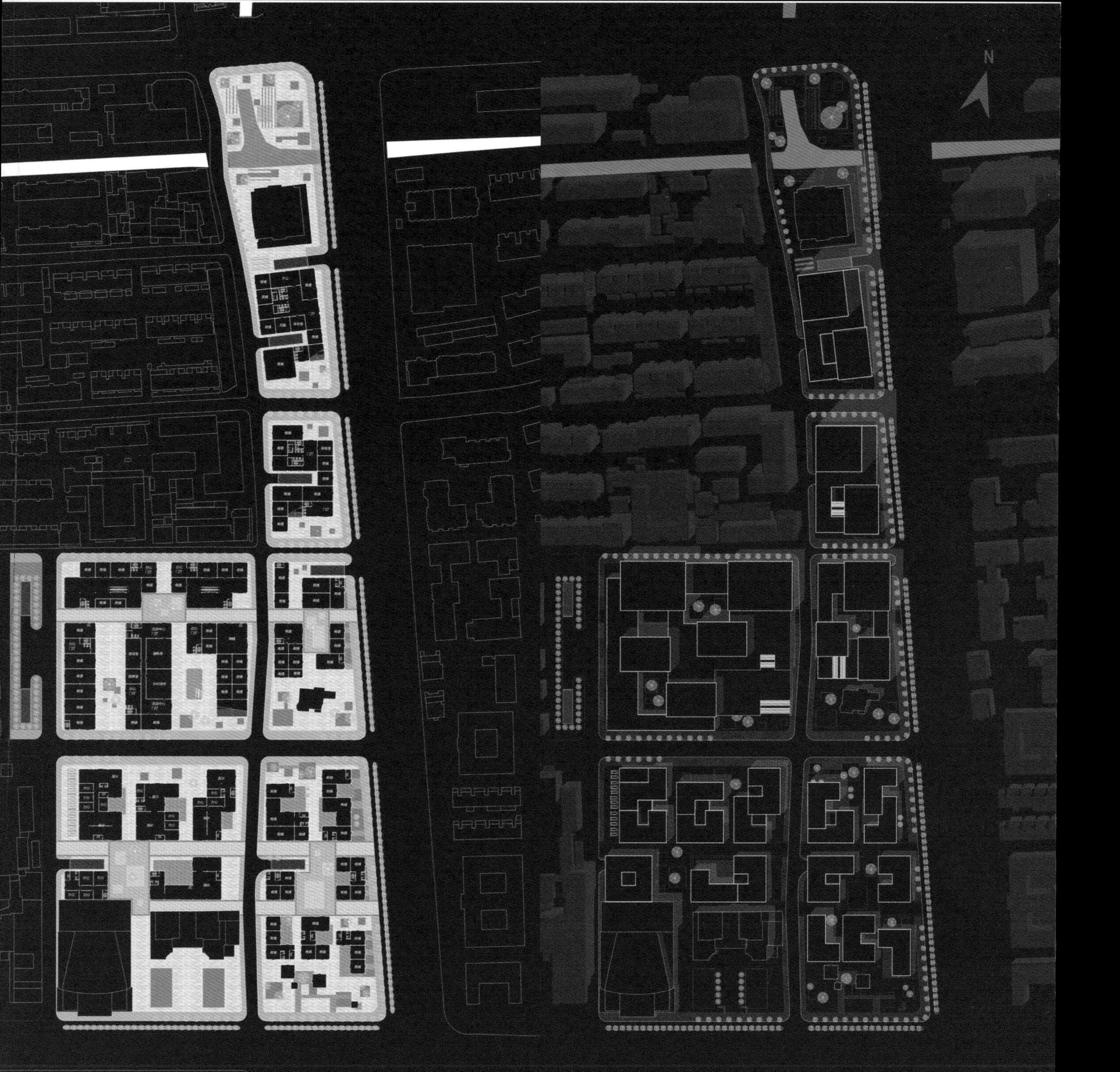
N

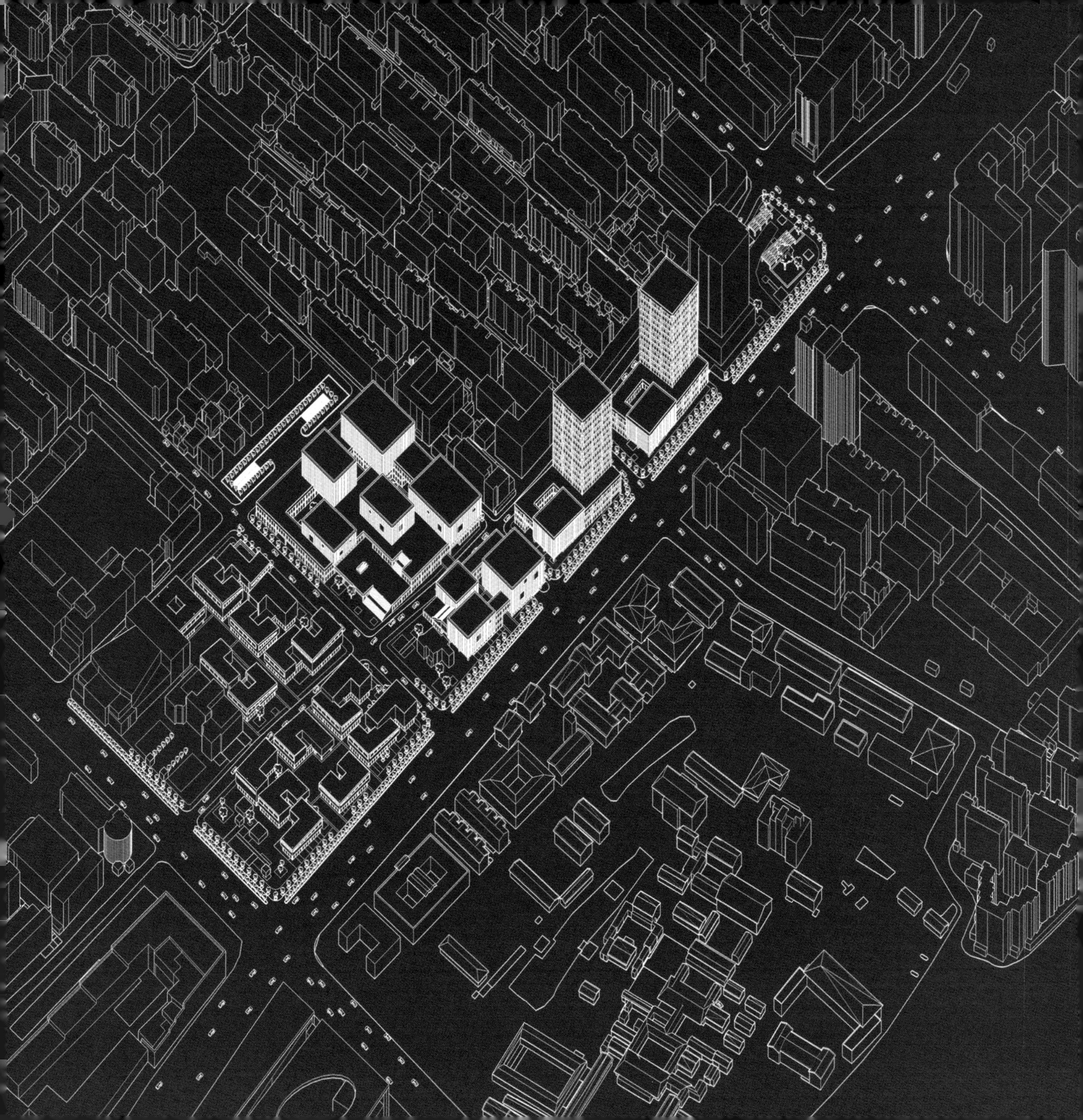

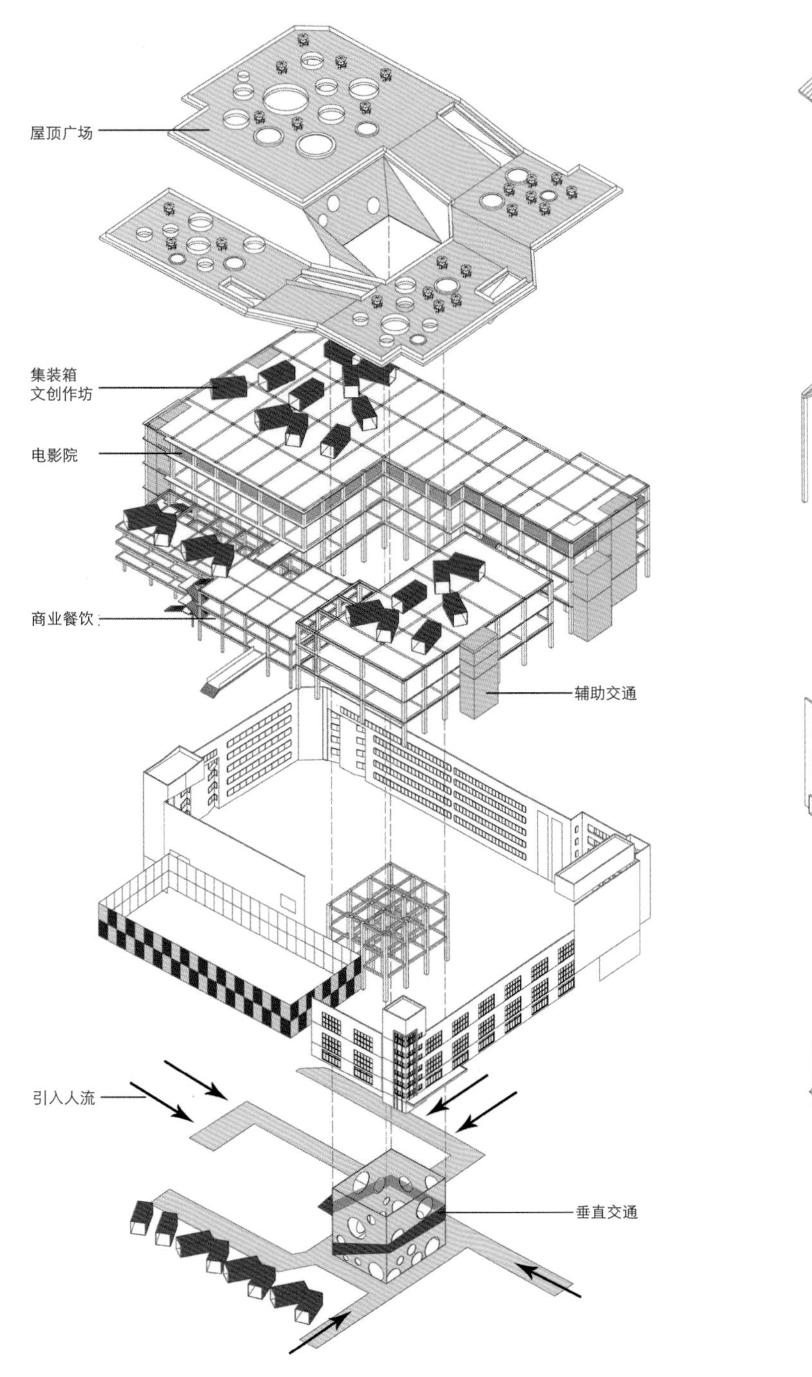

曼度文化广场改造示意

文化创意展示

店铺

坡顶厂房改造示意

WC
WC
WC
WC
WC
WC
Flo
WC
WC
WC
Floor 3
WC
WC
WC
WC

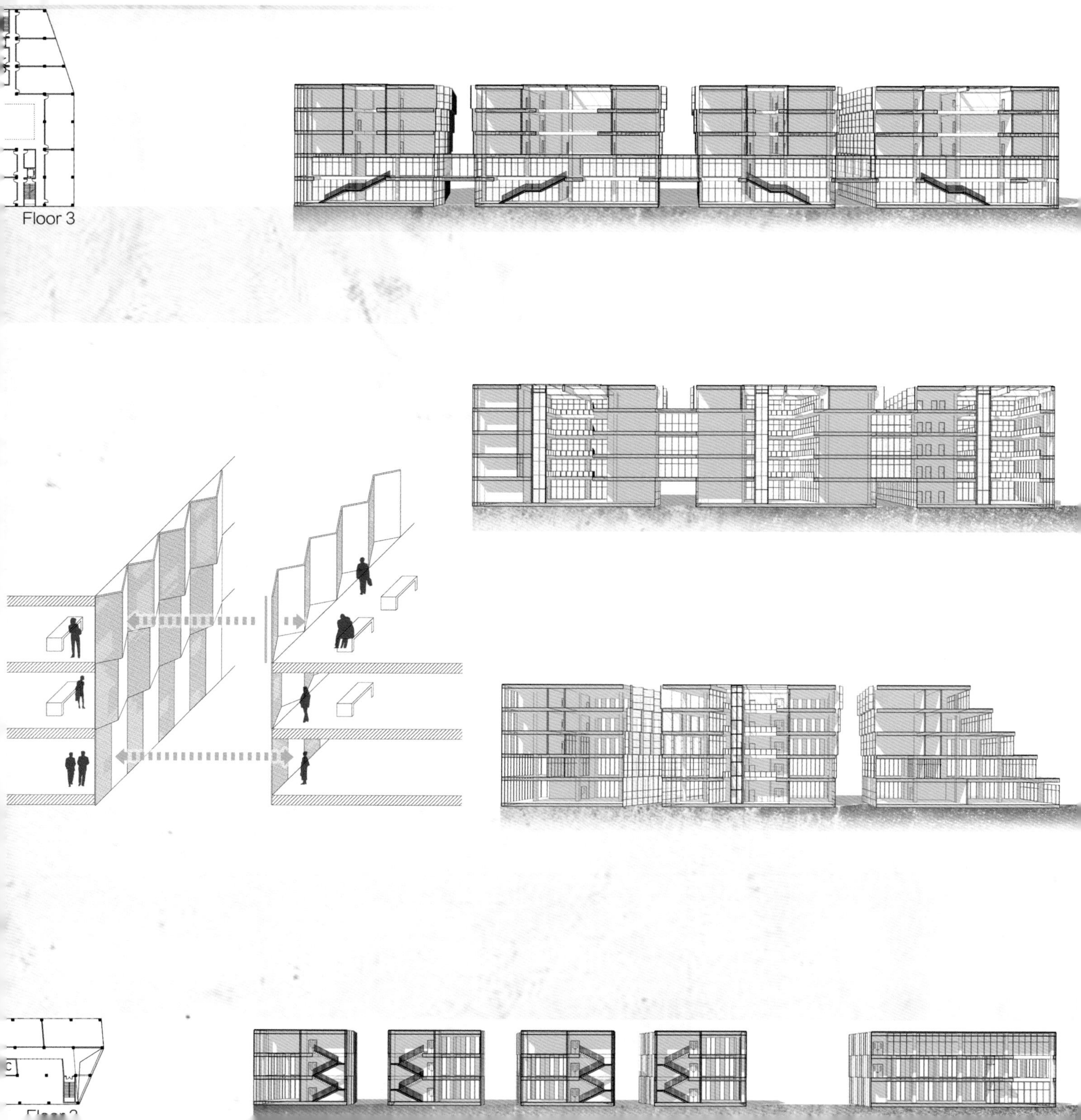
Floor 3

消费有逛公园的感觉，赞！
看！爸爸，有个商场！
走，我们进去看看
今天在民国博物馆学到了很多知识

引导学生思考这些场所可能承载的社会活动与身体行为，从而在身体尺度和个体经验上理解自己的设计。

Guide students to deliberate social activities and physical behaviors to be borne by such places, and to understand their design on basis of physical dimensions and personal experience.

弹性三维打印坐凳设计

DESIGN OF ELASTIC 3D PRINTED STOOL

钟华颖

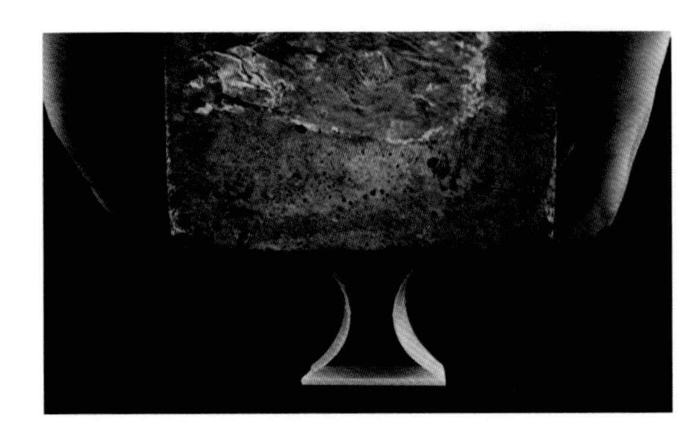

数字设计与建造作为毕业设计专题之一，在南京大学建筑与城市规划学院本科教学中已实行四年。开设这一专题的初始目的是为了完善从本科到硕士阶段的计算机辅助建筑设计教学体系，起到承上启下的作用。在总结学生本科阶段所学知识的同时，为研究生阶段继续深入学习做好思维和知识技能的储备。为实现这一教学目标，四年来教学计划根据上一年教学效果的反馈，从教学内容、教学进度安排、教学手段几个方面不断调整。今年的教学计划继续更新改进，通过制定分段教学目标并采用研究性的设计选题和数字模拟与实物搭建相结合等手段，建立标准化实验报告，初步实现了由强调数字设计知识技能的学习到注重数字设计思维培养的转变，完成了既定教学目标。

设计选题　今年的课程题目是弹性三维打印坐凳设计。依托学院最新的硬件条件，加工设备限定为三维打印机。将弹性三维打印作为研究对象是为了扭转三维打印仅用于复杂造型的认识，由仅关注形式转向材料性能，探索三维打印应用的可能性。坐凳则是以一个人体尺度的功能载体，串联本科阶段设计课所学基础知识。通过以上三个方面的限定，规范学生思考的方向和范围，尽快找到研究问题，展开设计研究。

教学内容　近年来数字设计教学发展迅速，已经成为建筑学教学的基本组成部分。相关知识在本科设计课教学所占比重也在增加。四年前课程开设之初，需要从基础知识开始授课的情况有了很大改变。今年教学中提供了基础知识的索引，内容由学生自学。实际的教学效果验证了自学掌握的知识点基本可以满足课程要求。

教学方法　根据往年教学的经验，学生容易模糊设计成果与科学问题，设计逻辑性和推导过程不清晰，设计进度缺乏控制。今年的教学借鉴了其他学科实验报告的方法，将每周的研究内容和设计成果以规范格式的实验报告形式进行记录。同时将研究问题进行分解，每周仅研究一个问题或者一个复杂问题的某一方面。对软件模拟到实物验证全过程的方法、过程、结果、数据进行记录。实验报告本身也成为毕业设计成果的重要组成部分，为下一届毕设提供了基础数据。

毕业设计的最终成果，基本实现了预期目标，打印出的坐凳原型可以承受人体重量，弹性变形提高了舒适度，验证了利用三维打印塑形方便的特性，可以通过合理的整体铸形获得一种不同于材料固有特性的新性能。设计成果参加了三维打印设计竞赛，参与申报双创成果展，体现了设计教学与科技创新结合的教学发展新趋势。

As one of the topics of graduation design, the course of digital design and building has been taught for four years in undergraduate teaching at the School of Architecture and Urban Planning, Nanjing University. This topic was established initially to improve the teaching system of computer-aided architectural design from undergraduate to graduate program, playing a linking role between these two stages. It provides reserve in terms of thinking mode and knowledge skills for further in-depth study in graduate stage while sums up knowledge already learnt during undergraduate years. To achieve this teaching objective, the teaching plan was adjusted constantly in those four years in aspects of teaching content, teaching scheduling, and teaching methods according to feedback of teaching effect of previous year. Update and improvement were also made in this year, standard experimental report was established through setup of phase-based objectives, and methods such as research-based topic selection, digital simulation and physical erection were integrated, which preliminarily realized the transition from learning that emphasizes knowledge and skills of digital design to the cultivation of thinking mode of digital design, and completed the set teaching objective.

Design topic selection　Course topic of this year is the design of elastic 3D printed stool. By relying on the latest hardware conditions of the school, the processing device is limited to a 3D printer. Using elastic 3D printing as research subject is intended to turn around the comprehension that 3D printing is only used for complicated molding, to turn attention from just forms to material properties, and to explore possibilities for application of 3D printing. The stool is a functional carrier based on dimensions of human body, and it is used to link fundamental knowledge learnt in undergraduate design courses. And limits in above three aspects specify direction and scope of the thinking by students, allow them to find research questions, and to carry out design research as soon as possible.

Teaching content　The teaching of digital design obtained rapid development in

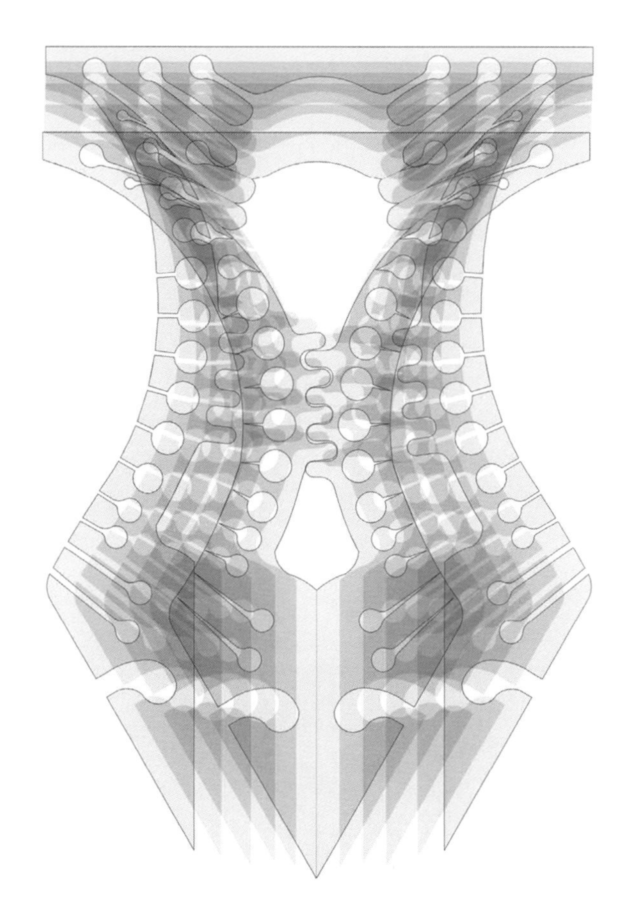

years, and has become a fundamental component of architectural education. The percentage of relevant knowledge in undergraduate design teaching has also been increased. In comparison with four years ago when this course was established, the situation requiring teaching from basic knowledge has been changed substantially. An index of basic knowledge was provided in this year's teaching process, which was studied by students themselves. The actual teach effect verifies that the knowledge grasped through self-study can satisfy the course requirements substantially.

Teaching methods. According to teaching experience from previous years, students tended to confuse design results and scientific issues, with unclear design logic and deduction process, and were lack of design progress control. Methodology of experimental report was borrowed from other disciplines in this year's teaching process, and research content and design results were recorded weekly in experimental report of standard format. Meanwhile, the research issue was broken down, and research was done for only one issue or one aspect of a complicated issue in one week. Methods, process, results, and data of the entire process from software simulation to physical verification were recorded. Experimental reports are also an important component of results of graduation design, offering basic data for graduation design by students in next year.

The final results of graduation design realized the expected objective substantially. The printed stool prototype can bear the load of a human body, and the elastic deformation increases the level of comfort. It verifies the convenient feature of molding with 3D printing, and that a new property different from inherent features of material may be obtained through reasonable overall molding. The design results took part in a 3D printing design contest, applied for double-innovation achievement exhibition, and reflected the new development trend of teaching that combines design teaching with scientific and technological innovation.

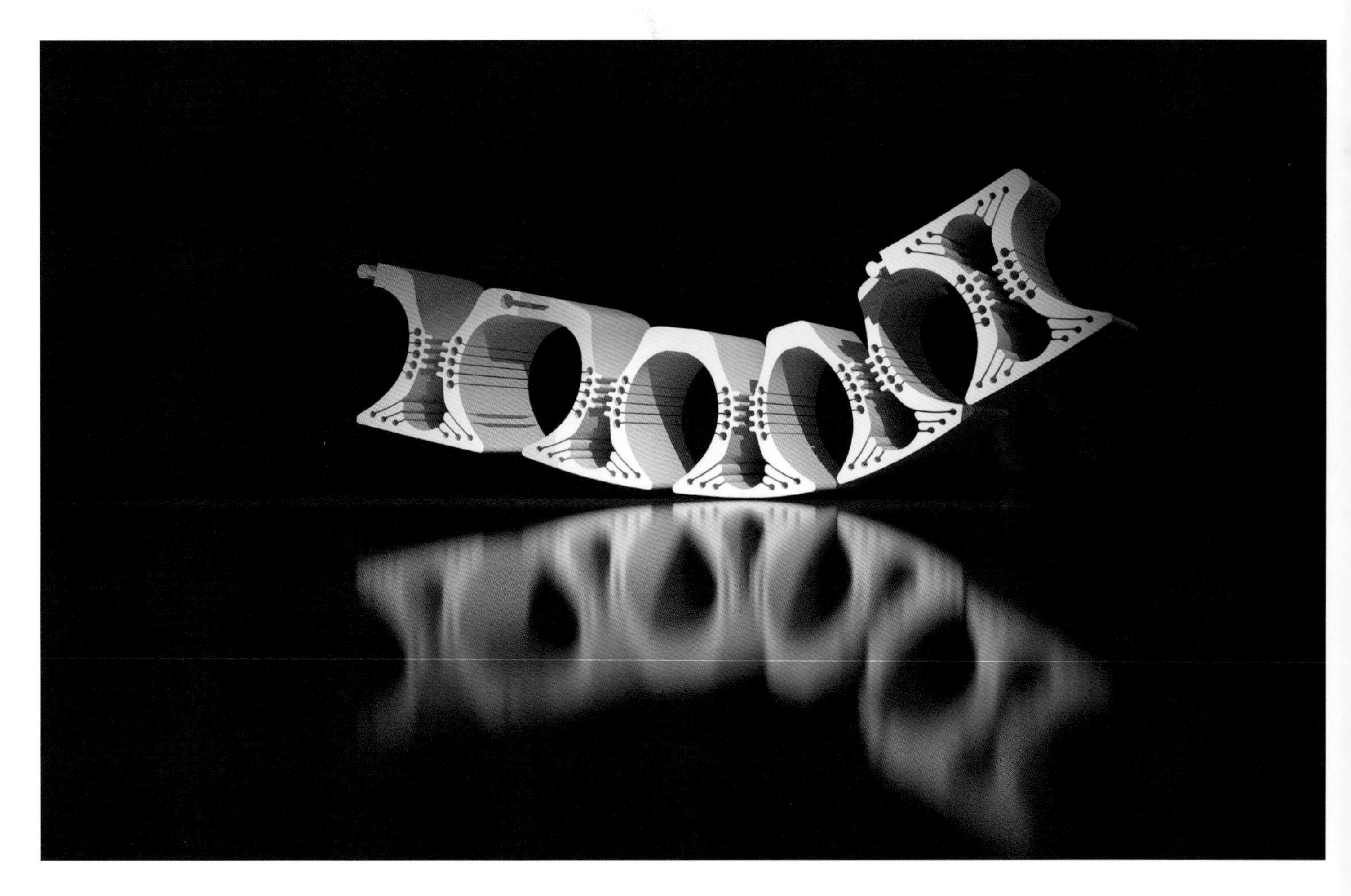

采用插槽式的连接方式，可以在保证稳固的情况下，增大用具的变形能力，从而增加舒适度，减小连接端受到的剪力，减小了断裂的可能性。

The stool adopts slot-type connection, which ensures the stability while improving the instrument's deformability, thus improving the comfort level, decreasing the shearing force on the connection end and reducing the possibility of fracture.

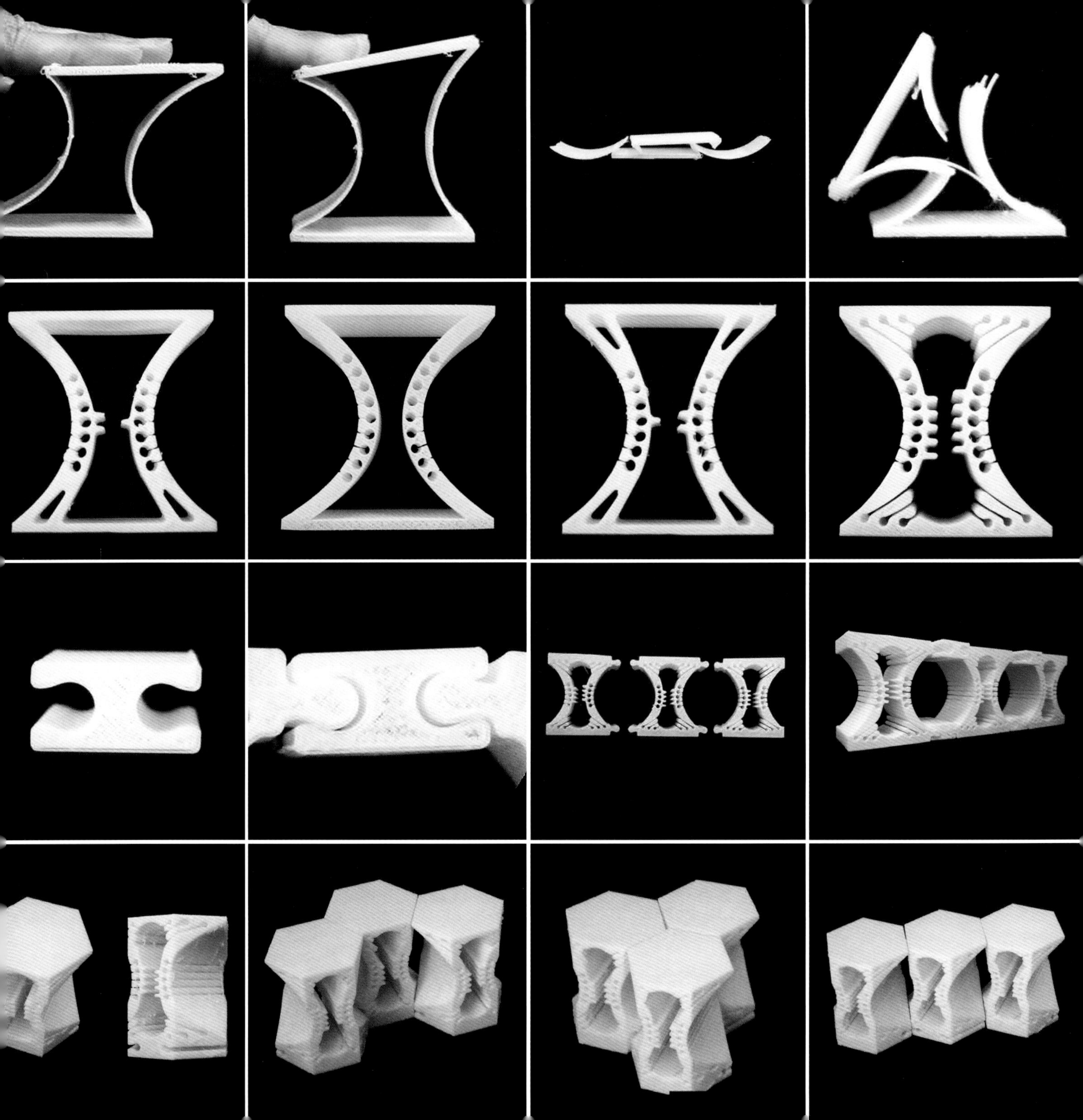

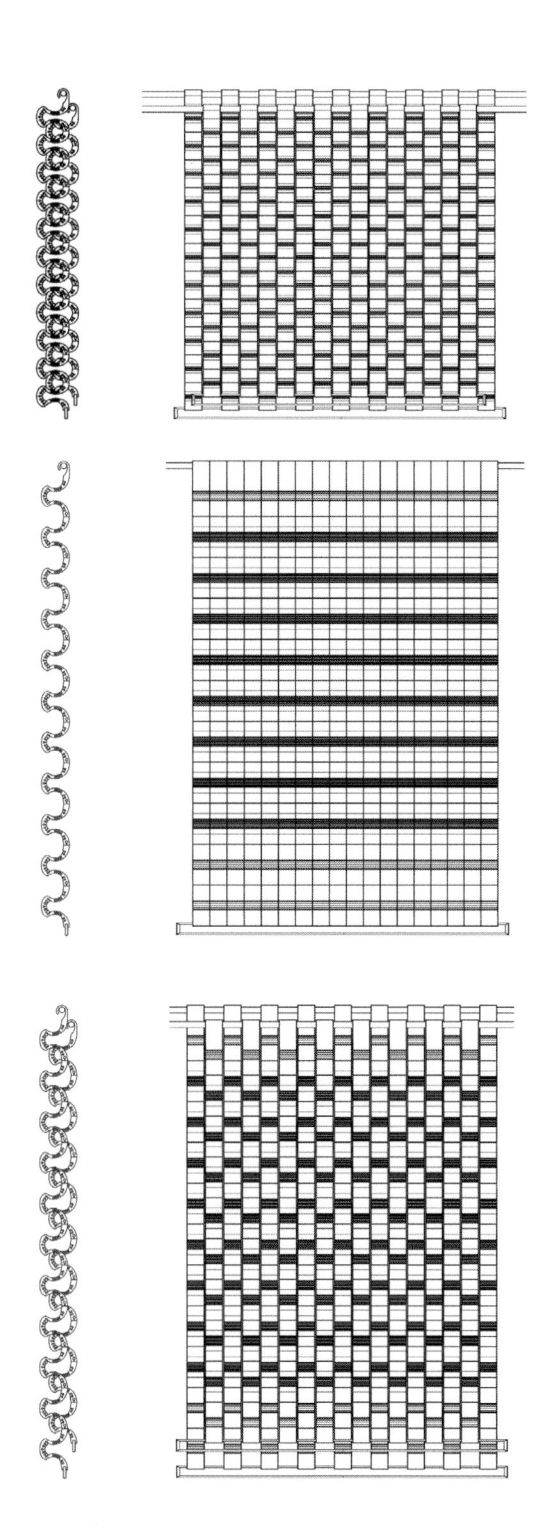

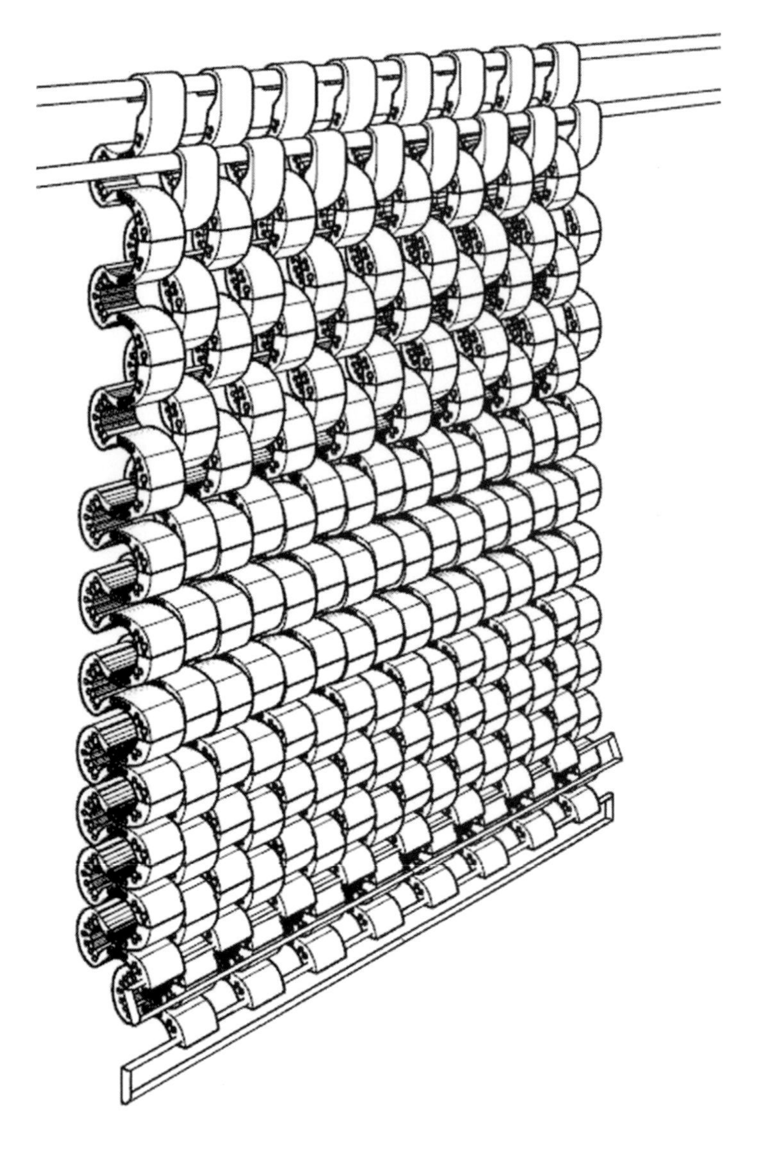

基本单元组合成遮阳构件，将其拉下并固定能起到良好的遮阳效果，而由于材料本身的特性，也能透过温和的柔光。在需要一定光线时可以将构件交错排列，达到类百叶窗的效果。

The basic units are integrated into sunshade components, which have good shading effects if pulled down and fixed, and allow transmitting of soft light due to the property of the material. The components can be made in stagger arrangement to adjust the brightness of light, thus realizing the effects similar to shutters.

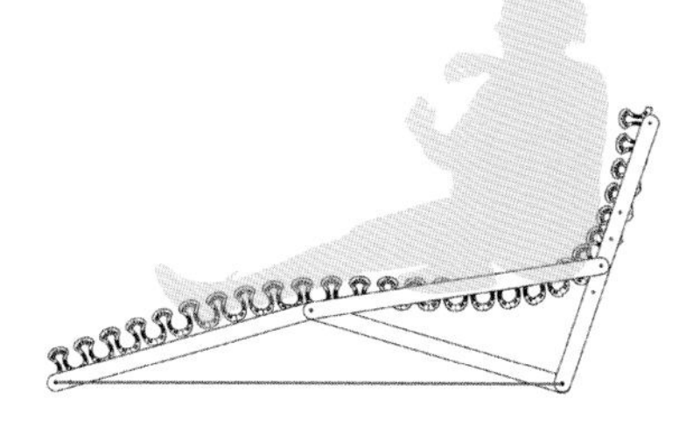

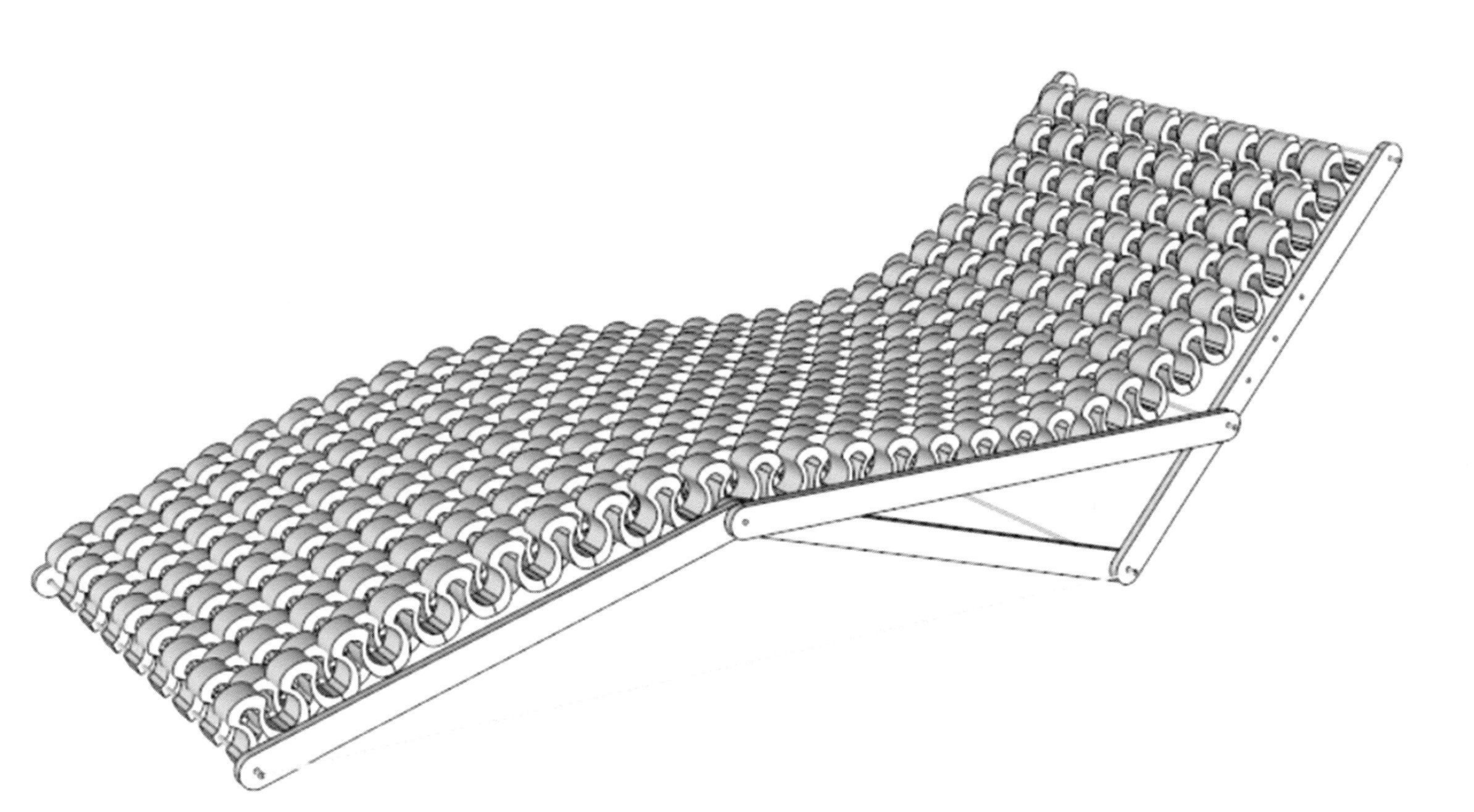

基本单元拼接成条带状，交错叠加覆盖在支架上，组合成躺椅。通过调节支架以满足各种需求，使体验更多样化。

The basic units are spliced into strips, overlaying the support in a stagger way, to form a deck chair. It can meet various demands and realize varied experience by adjusting the support.

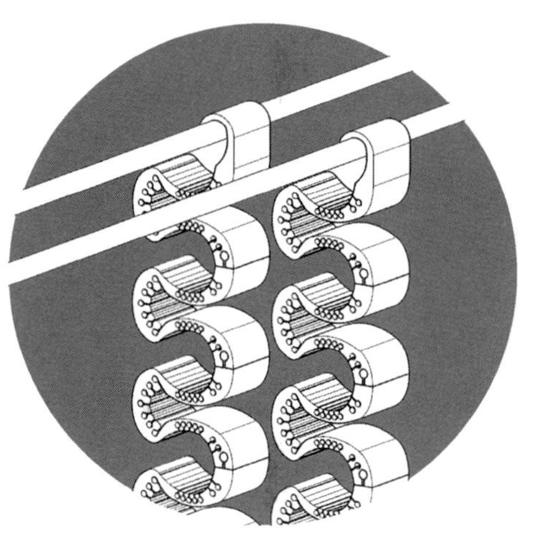

长汀历史名城更新与建筑设计

RENEWAL AND ARCHITECTURAL DESIGN OF CHANGTING HISTORICAL CITY

丁沃沃 胡友培

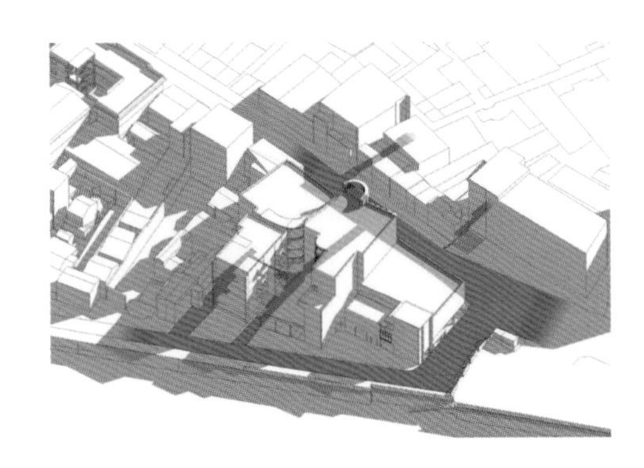

城市更新是现阶段我国城市化面临的重要任务，同时也是城市发展的机遇与难点。任何常规和传统的思维方式、设计手法甚或规范，都面临着不可回避的挑战。为了让建筑学本科学生在完成学业之前尽可能地接触到现实问题，领会国家需求并体验到实际应用中对专业知识的需求，我们以“福建长汀县历史文化名城保护与更新的城市设计”这个真实项目为基础，通过对真实场地的认知和对专业文本的研读，了解古城的物质空间特征、现实状况、城市需求和改造难点。作为毕业设计，并不需要模拟现实中的项目设计，而是鼓励学生基于现实条件和难点，尝试通过设计创意解决问题，切实体验到如何运用设计知识解决现实问题。

长汀县位于福建省西南部闽赣边境，依卧龙山而傍汀江，城内保存了众多的寺庙、祠堂和传统大宅院，既有众多的传统木构建筑和夯土建筑，又有民国时期的闽南洋房，1994年被评为国家历史文化名城。历史上，长汀县被称为客家首府，是客家文化重要的聚集地。同时，长汀又是重要的红色根据地，中华人民共和国建国时期的主要领导人都曾在此地逗留或小住，古城内保留着多处红色文化遗址。古城周围群山环绕，所以又被评为国家级生态城市。虽然古城拥有独特的旅游资源，但是近年来经济发展的诉求使古城面临着巨大的压力。保护古城和古建筑早已不只是技术的问题，单纯的保护早已使城市不堪重负，需要通过创新规划与保护的思路，创新建筑设计方法才能找到城市更新的出路。作为毕业设计的案例，它不仅提供了丰富的历史文化知识，而且也提供了各种优秀的民居类型、地方材料和通用做法，这些都成为毕业设计丰富且宝贵的资源。

我们认为本科毕业设计应该强调研究型设计而不是建筑专业设计模拟。因此，毕业设计的全过程应该体现以设计研究为主线，毕业设计的各个过程应该按设计研究的路径设置而不是按项目规律设置，更为重要的是这个主导思想还应该贯彻在毕业设计的各个阶段之中。为此，该课题首先设置了5个设计研究的问题：如何调研方能挖掘真正的问题？如何将繁杂的民居归类？如何理解当地材料和工法？如何理解保护更新建筑设计的问题？如何理解建筑设计与设计方法？

调研：作为从未接触过研究的本科生，毕业设计小组在出发前首先接受了城市设计调研和资料收集的训练课程，带着问题深入实际。通过调研理解了古城的总体构架空间结构、交通体系以及风貌特征，提高了本科学生一般都缺乏的宏观意识。其次，基于收集的资料，学生整理出街巷结构、建筑类型和细部做法，对于古城的魅力所在形成了较为清晰的认知，为进一步的设计奠定了基础。更为可贵的是，通过调研学生们对人与城市空间的关系以及建筑不是一个在建筑师手中把玩的物体，而是与人们生活息息相关的物质空间有了进一步的认识与体会，并认识到了建筑的社会意义和价值。总之，具有一定方法论基础的调查与研究，帮助学生更为全面而深刻地理解建筑现象与设计活动。根据地块特征和城市更新的一般问题，设计课题选择三个代表性地块作为设计场地：

A 地块——以文保建筑为主的地块。该地块中现存各种等级的文保建筑，其中包含周恩来和刘少奇在长汀工作时的居所。如何挖掘该地块的历史价值、如何展示和利用该地块的历史价值成为场地设计和建筑设计的主要命题。

B 地块——公产权和私产权交织的商业地块。该地块地处商业街中段，有良好的商业价值，然而现有建筑状况十分破败。由于商业运行模式陈旧，商业价值没有充分开发，地块内有价值的传统建筑也没有得到很好的展示和利用。因此，如何解决复杂条件地块，挖掘和开发商业空间，保护和延续传统街道风貌成为该地块需要讨论的问题。

C 地块——城市节点的重要地块。该地块位于老城重要城门之一济川门外，同时又是重要商业街的入口。该地块面迎客家传统风貌的古城，背对民国风貌的商业街，同时作为现代城市的一部分该地块周边交通复杂而繁忙。该地块在过去的城市发展中，已经因需要建设了 5~6 层的高大建筑，然而质量较差，因功能早已不适应而成为空楼。从城市的角度看，这是一块价值极高的地块，周边繁杂拥挤的商业呼唤新的可利用的空间，地块开发存在最大化的容积率诉求；从老城保护的角度，过大的建筑体量会损害风貌的保护与协调。因此，如何通过设计创新，解决功能、交通和形式问题，协调开发与保护的冲突，成为该地块的设计焦点。

分析：我们认为研究型设计主角依然是设计，研究型设计和纯粹学术研究的最大区别不在于是否调研，而在于如何看待调研资料，因此如何分析，即如何看待手中的资料是关键。为此，学生花了两到三周的时间，基于设计问题对调研资料展开分析，即分析的结果要回答设计的问题所在、设计的难点、设计的可能性或设计的突破点。此时，要杜绝一般性分析、完全不针对问题的分析、没有结论的分析或不解决建筑设计问题的结论。严格的分析训练，不仅提高了学生的研究能力和思维逻辑，而且有助于学生形成问题导向的设计意识：用设计解决问题。要完成这个任务，分析图的表达非常重要，在训练中强调了分析图不仅要展现设计的“空间”和潜在的“可能性”。

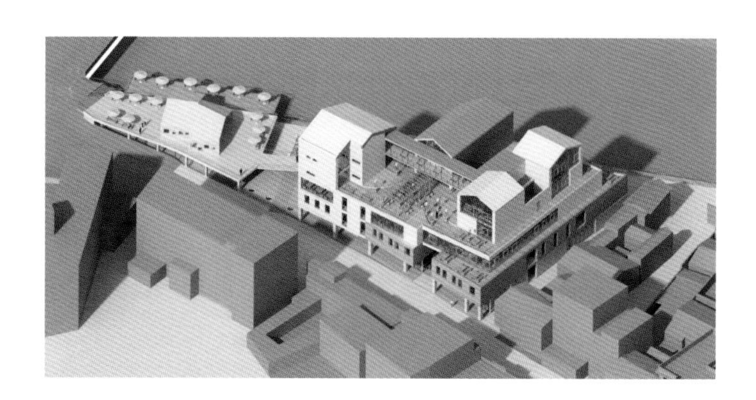

设计：所谓设计创意是通过创造性的设计漂亮地解决问题，研究型设计的意义正是在于探索设计新思路或新方法。因此，我们认为学习研究型设计是培养设计创意思维的基础。区别于常规的设计方法，研究型设计首先强调的是设计思想。具体而言，就是强调设计概念在引领整个设计过程中的重要地位。由于缺乏基础，本环节的训练延续的时间比原设定的时间要长得多，但是通过该阶段的训练，学生在思维水平上得到较大的提升。这里需要说明的是，概念设计的训练一定要和设计表达相结合，虽然概念设计强调研究性，但是概念设计的表达需要能表达概念的图纸和具体的结合场地的空间分配。最终，由概念引领了空间布局、形态决策、形式策略以及细部处理，直至成果表达和答辩表述。

学生通过整个毕业设计环节，领略了设计研究的价值和意义。设计成果分为三大部分：调研文本、测绘文本、设计文本。

结语：本次毕业设计涵盖历史文化知识、典型民居类型、建筑设计与建造以及城市设计方法等训练计划，旨在通过训练学习科学有效的调研与分析方法、多重限定下的建筑设计，理解面向建造的真实问题和城市设计的现实意义。在将本科所学知识融会贯通的基础上，理解设计与研究的关系和研究对于设计的价值，并通过具体的操作与练习，转化为个体化的知识与技能。

Urban renewal is an important task faced by current urbanization in China, and also an opportunity and challenge for urban development. All regular and traditional thinking modes, design methods, and even standards are confronted with inevitable challenges. To enable undergraduate students of architecture to come into contact with practical problems to the greatest extent as possible before graduation, comprehend national demand, and experience the need of professional knowledge in practical applications, we are based on the practical project of "Urban Design for Preservation and Renewal of the Historic and Culture City of Changting County in Fujian Province" to get to know characteristics of physical space, realistic conditions, urban demand, and challenges for renewal of the ancient city through perception on real-world site and study on professional literature. As a project of graduation design, it does not require simulation of project design in real world, but encourages students to attempt to solve problems with design innovation, and experience the way of solving practical problems with design knowledge on basis of practical conditions and challenges.

The Changting County is located at the border between Fujian and Jiangxi in southeast of Fujian Province, near Wolong mountain and by Tingjiang river. A large number of temples, ancestral halls, and traditional compounds are preserved in the city. There are both traditional wooden-structure buildings and rammed earth buildings, as well as western-style houses with southern Fujian characteristics that were built in the period of Republic of China, and it received the honor of a national historic and culture city in 1994. In history, Changting County was known as the capital of Hakkas, and was an important gathering place of Hakka culture. Meanwhile, Changting was also a red revolutionary base, primary leaders were stayed or lived here for a while in the process of founding the PRC, and there are still several red cultural remains preserved in the ancient city. The ancient city is surrounded by mountains, thus it is appraised as a national ecological city. Although the ancient city has unique tourist resources, it faces huge pressure in term of economic development in recent years. Preservation of the ancient city and its ancient building has not been a technical issue for a long time, pure preservation has long made the city overloaded, it requires ideas for innovative planning and preservation, and only innovative architectural design method can find a way for renewal of the city. As a case for graduation design, it provides not only abundant knowledge about history and culture, but also various types of excellent folk dwellings, local materials, and common practices, and all these are rich and valuable resources for graduation design.

In our opinion, graduation design for undergraduate students shall emphasize research-based design instead of simulation of professional architectural design. Therefore, the overall process of graduation design shall reflect the main line of design research, and various processes of graduation design shall be arranged on basis of the path of research, instead of rules of project, and more importantly, this guiding thought shall be followed in various stages of graduation design.

Therefore, five questions about design research are listed for the task: How should we do investigation to dig real problems? How should we classify the complex and diversified folk dwellings?How could we understand local materials and construction methods?How should we understand the question of architectural design for preservation and renewal?How should we understand architectural design and design methods?

Investigation: As undergraduate students who never contact with research before, the group of graduation design first received a training course about investigation and data collection for urban design before starting off, and went deeply into reality with questions. They understood the structure of overall framework space, traffic system, and landscape features of the ancient city through investigation, and improved their macro-awareness that are lack of generally for undergraduate students. And secondly, students worked out structure of streets and lanes, building types and detail construction method on basis of collected data, and had a relatively clear perception to the charm of ancient city, which laid down foundation for further design work. And more valuably, students acquired further understanding and experience on relations between human and urban space through investigation, and that building is not an object being manipulated by the architect, but physical space that is closely connected to human life, and realized the social meaning and value of architecture. In a word, investigation and research on basis of certain methodology enabled students to understand architectural phenomenon and design activities in a more complete and profound manner. On basis of land parcel characteristics and general issues of urban renewal, three representative land parcels are selected as design sites for the design task:

Land parcel A—the one mainly of protected cultural relic buildings. There are various levels of protected cultural relic buildings within this land parcel, including residence of Zhou Enlai and Liu Shaoqi when they worked in Changting. The way to excavate historic value of the land parcel, and the way to display and utilize historic value of the land parcel are main topics for the site design and architectural design.

Land parcel B—a commercial land parcel with mixed public and private properties. This land parcel is located at central section of a commercial street, with good commercial value, but existing buildings here are highly dilapidated. Due to outdated business operation mode, its commercial value has not been fully exploited, and valuable traditional buildings within the land parcel have also not been properly displayed and utilized yet. Therefore, the way to work out a solution for the land parcel with complicated conditions, to excavate and develop commercial space, and to protect and continue traditional street features are main issues to be discussed for the land parcel.

Land parcel C—an important land parcel of urban nodes. This land parcel is located outside Jichuan Gate – one of the important gates of the old town, and also an important entrance to the commercial street. The land parcel faces to the old town of traditional Hakka features, and is back against the commercial street of features of the Republic of China era, and as part of modern city, the traffic network around it is complicated and busy. During urban development in the past, high buildings with 5~6 floors were built within the land parcel due to its demand, but given the poor quality and outdated functions, they have become empty buildings. From the perspective of a city, this is a land parcel of extremely high value, the diversified and crowed businesses around it call for new available spaces, and development of the land parcel appeals for maximized floor space ratio; and in the view of old town protection, oversized building mass will damage the preservation and coordination of features. Therefore, the way to solve problems in connection with functions, traffic, and forms and to coordinate the conflict between development and protection through design innovation has become a focus of design for the land parcel.

Analysis: In our opinion, the main role of research-based design is still design, the biggest difference between research-based design and pure academic design does not lie in whether investigation is done or not, but the way to view investigation data, and the way of analysis, that is, the key is the way how do we view data in hands. To this end, students spent two to three weeks on analysis of investigation data based on design issues, that is, results of analysis should answer questions of the design, difficulties of design, probability of design or breath-through point of design. Now the general analysis, analysis not targeting at any problem at all, analysis without conclusions or conclusions not solving architectural design issues must be eliminated. Strict analysis training not only improved research ability and thinking logic of students, but also helped them shape problem-oriented design awareness: solving problems with design. To complete this task, the presentation of analysis chart is very important, and the training emphasized that analysis chart should present “space” of design and potential “possibilities”.

Design: The so-called design creativity is to solve problems beautifully through creative design, and the significance of research-based design is just exploring new design ideas or new methods. Therefore, we believe that learning research-based design is the foundation for cultivation of creative thinking of design. Different from regular design method, what emphasized first by research-based design is the design thought. Specifically, it emphasizes the important position of design concept in guiding the entire design process. Training of this link lasted much longer than the duration expected, but thinking capacity of students was improved significantly through training in this phase. It is worth noting that training of conceptual design must be combined with design expression. Although conceptual design emphasizes the nature of research, expression of conceptual design requires drawings that can express concepts and specific space allocation in combination with the site. And in the end, concepts lead to spatial arrangement, morphological decision, form-oriented strategy, and treatment of details, and finally to result expression and defense presentation.

Students realized the value and significance of design research through the entire process of graduation design. Design results consist of three major parts: investigation documents, surveying document, and design documents.

Conclusion: This graduation design covers historic and culture knowledge, typical types of folk dwellings, architectural design and construction, as well as urban design method and other training plans. It aims to understand construction-oriented real-world questions and practical significance of urban design through training and learning of scientific and effective investigation and analysis methods, and architectural design under multiple restrictions, to understand relations between design and research and the value of research to design on basis of digestion and connection of knowledge learnt in undergraduate courses, and to transform it into personal knowledge and skills through specific operation and exercise.

一号地块
宋慈路
二号地块
水东街
三号地块

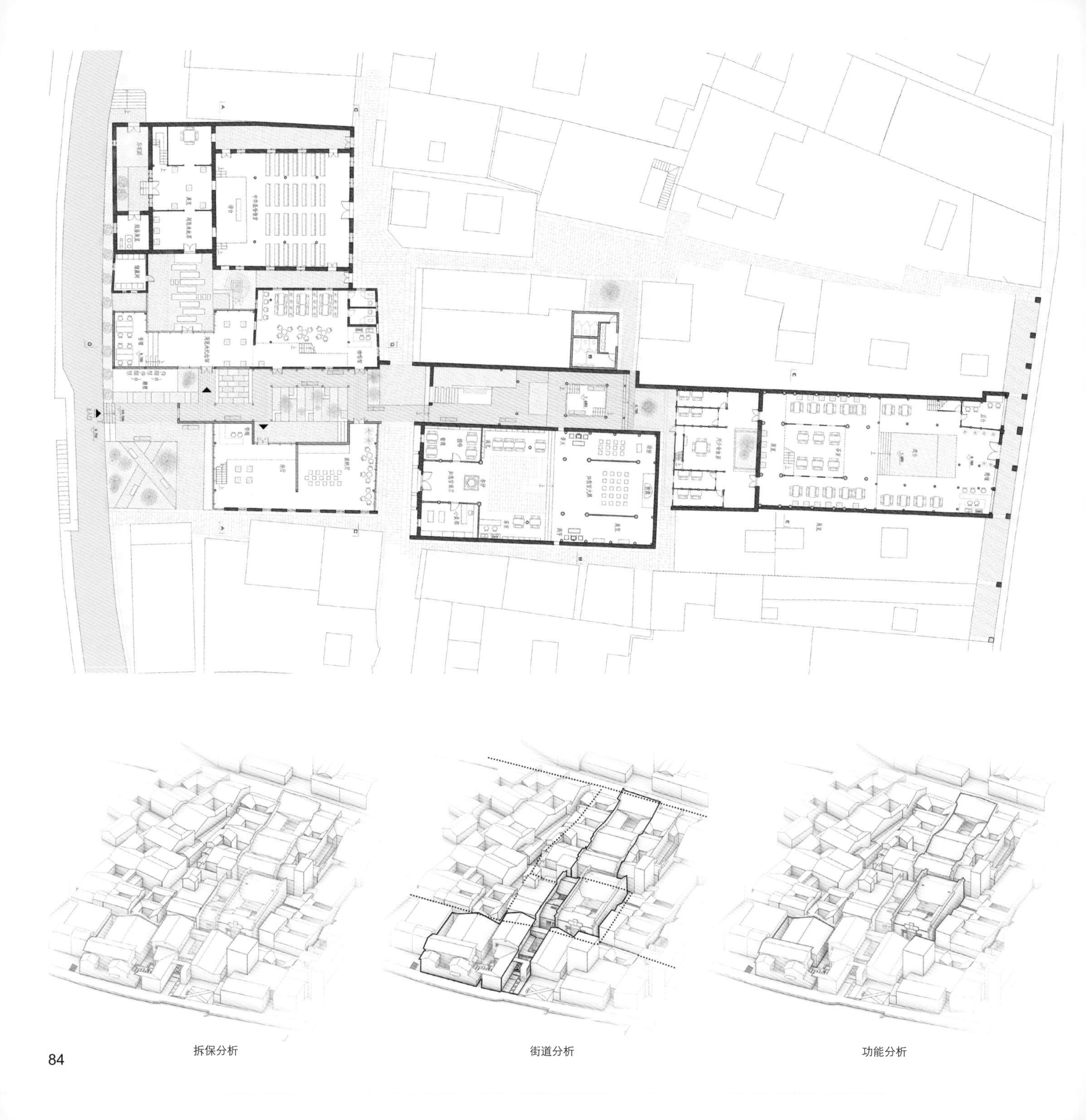

拆保分析

街道分析

功能分析

通过调研，整理出街巷结构、建筑类型和细部做法，对人与城市空间的关系有了进一步认识与体会。

Students worked out structure of streets and lanes, building types and detail construction method, and acquired further understanding and experience on relations between human and urban space through investigation.

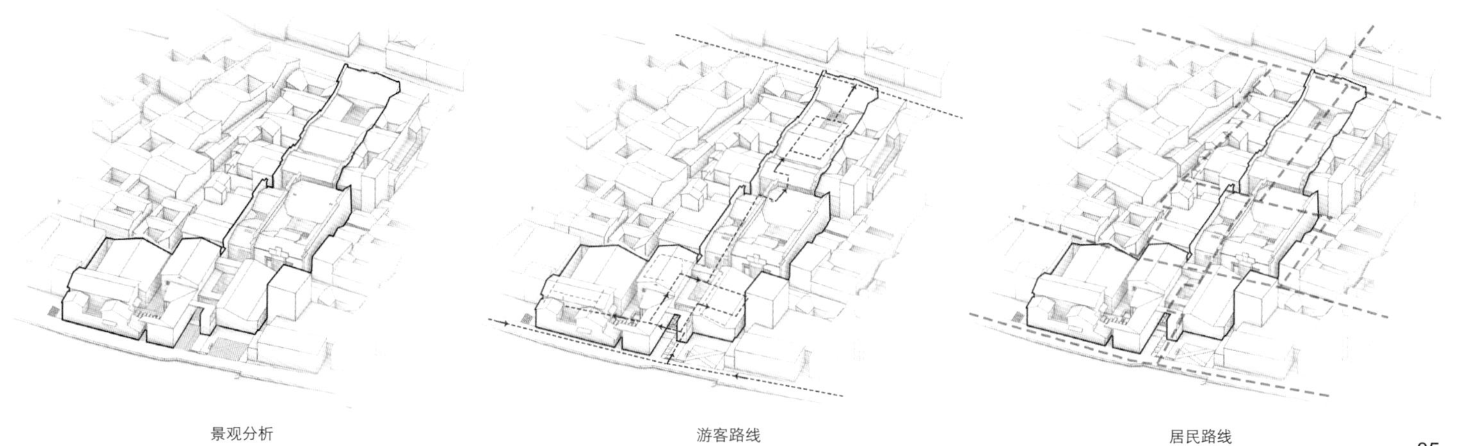

景观分析　　游客路线　　居民路线

以“红色记忆中的生活”为概念，改造打通仙隐观，设计一条可以容纳居民休闲交流的街道，连接两座红色故居，让生活成为展览的一部分，让居民与游客同时感受场地与红色记忆。

Renovate and get through to Xianyin Guan on basis of the concept of “life in red memories”, design one street available for relaxation and exchange by residents to connect two “red” former residences, make life part of exhibition, and allow residents and tourists to feel both the site and red memories.

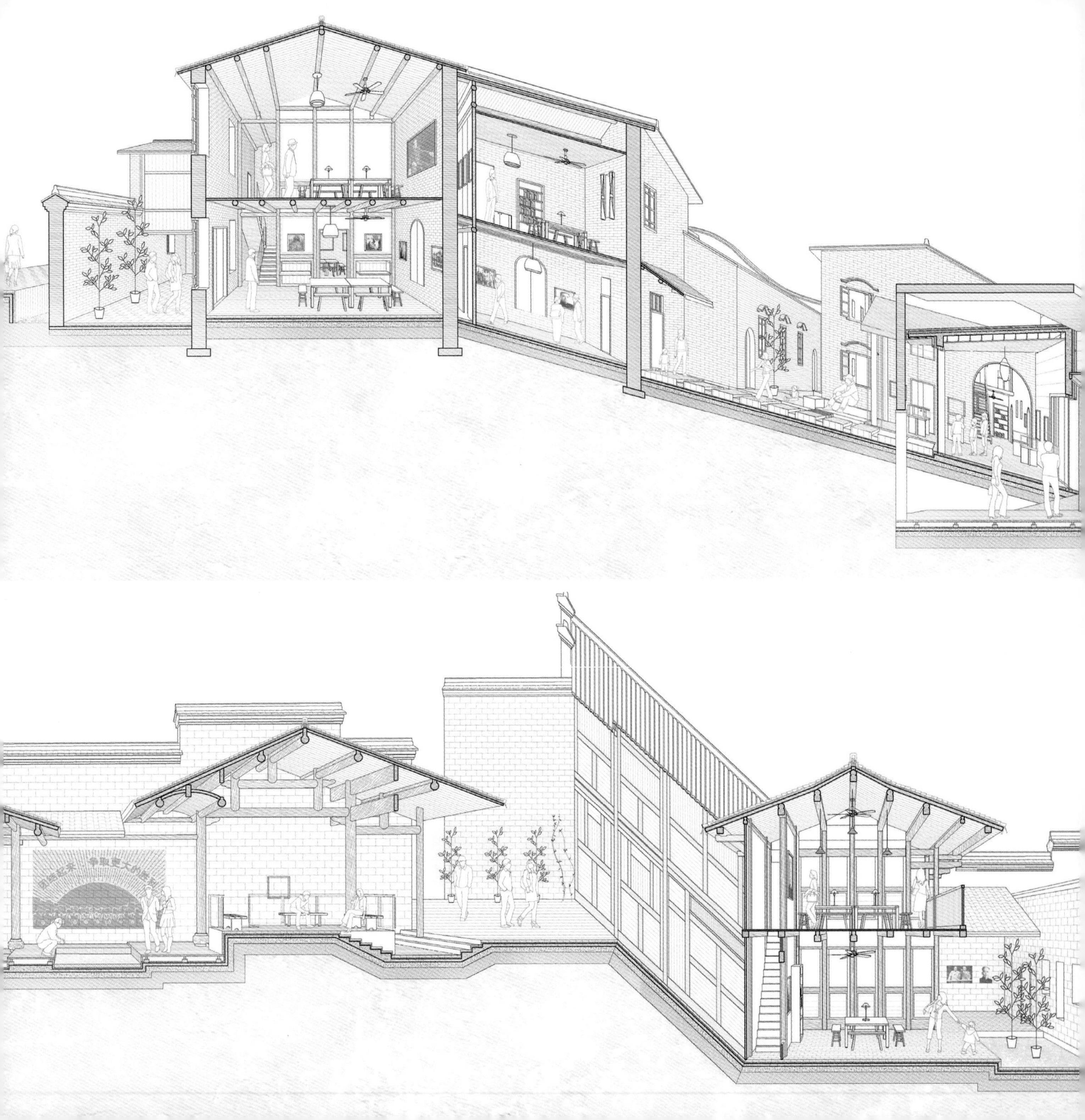

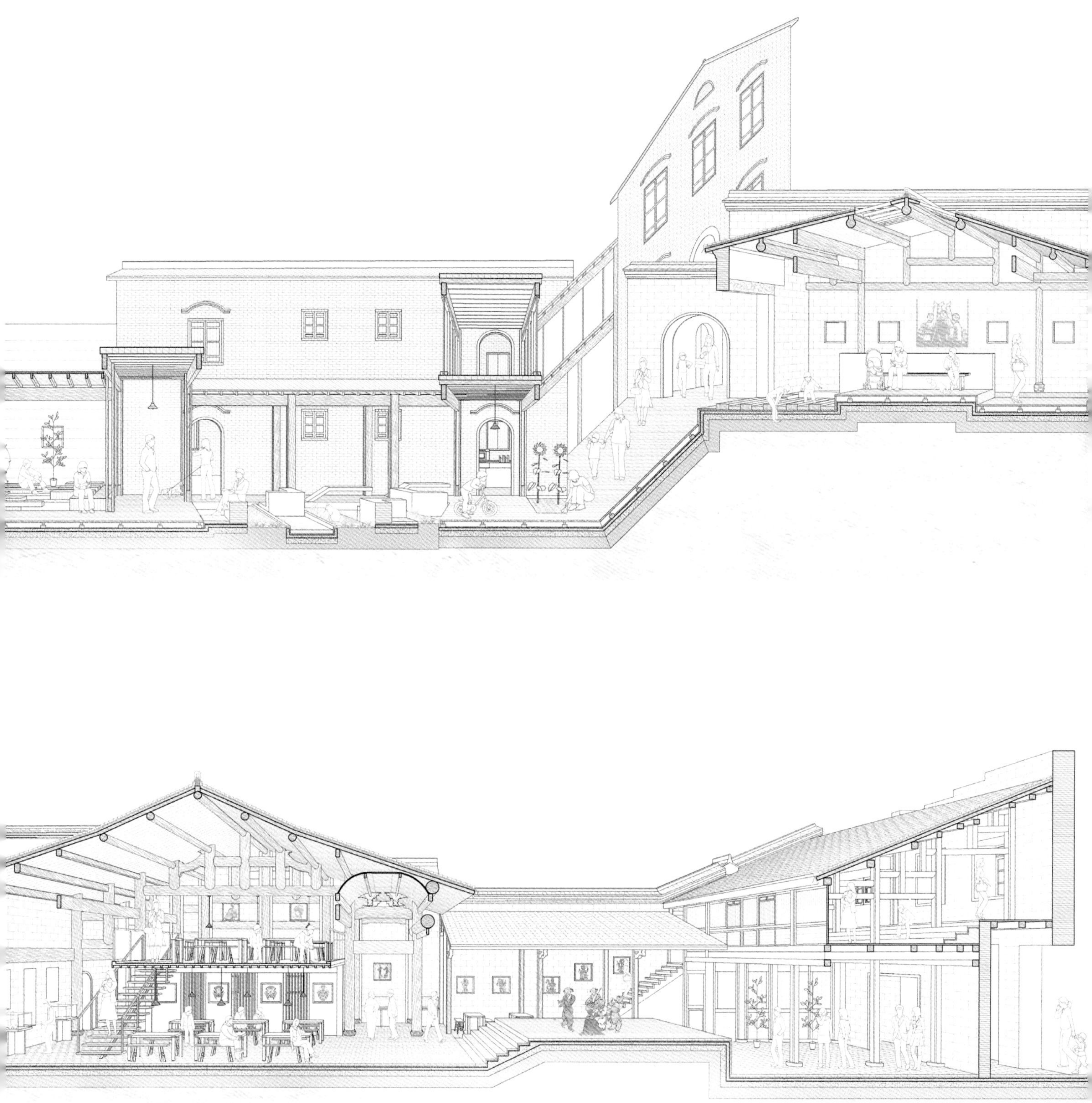

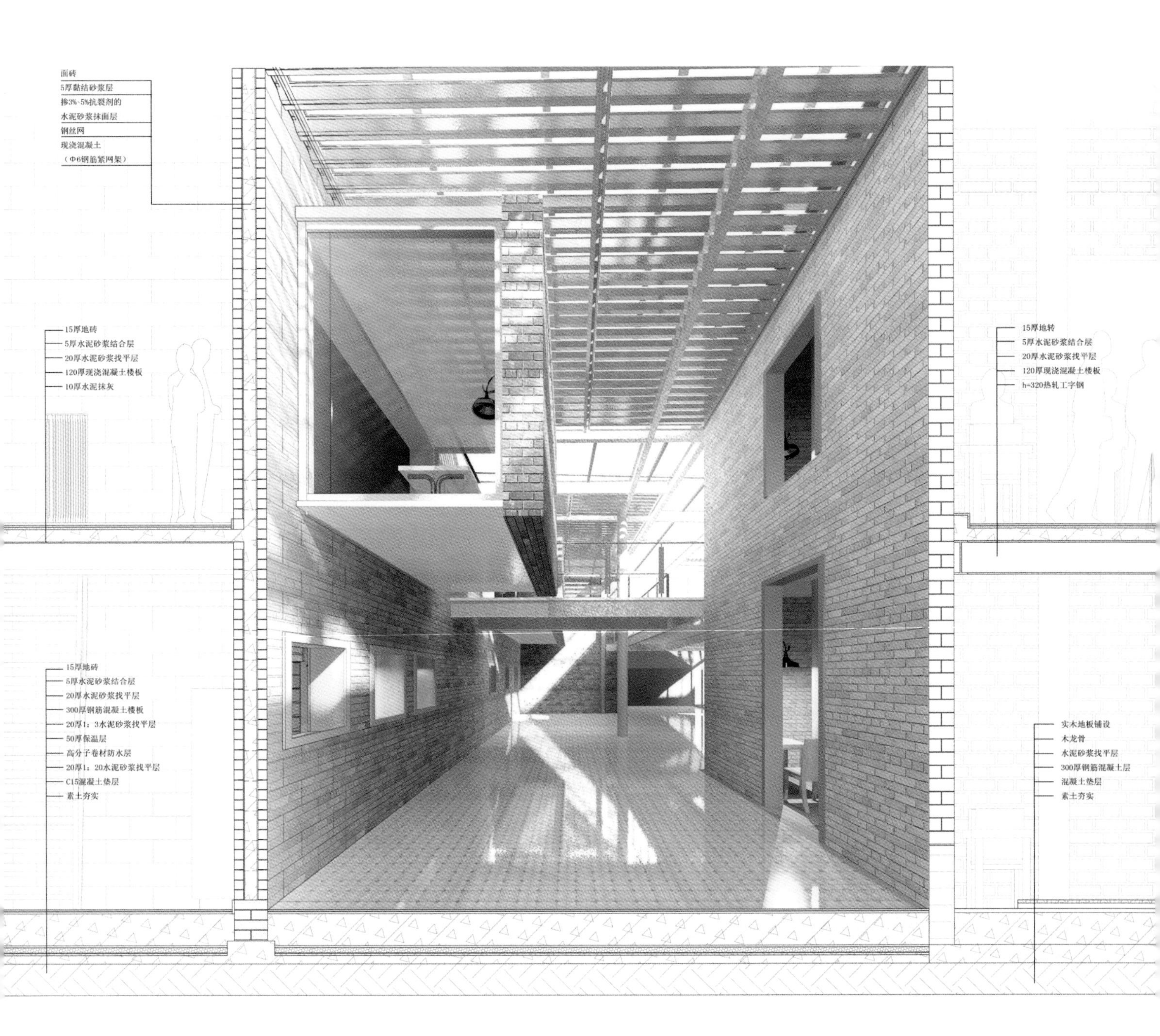
面砖
5厚黏结砂浆层
掺3%-5%抗裂剂的
水泥砂浆抹面层
钢丝网
现浇混凝土
（Φ6钢筋紧网架）
15厚地砖
5厚水泥砂浆结合层
20厚水泥砂浆找平层
120厚现浇混凝土楼板
10厚水泥抹灰
15厚地转
5厚水泥砂浆结合层
20厚水泥砂浆找平层
120厚现浇混凝土楼板
h=320热轧工字钢
15厚地砖
5厚水泥砂浆结合层
20厚水泥砂浆找平层
300厚钢筋混凝土楼板
20厚1：3水泥砂浆找平层
50厚保温层
高分子卷材防水层
20厚1：20水泥砂浆找平层
C15混凝土垫层
素土夯实
实木地板铺设
木龙骨
水泥砂浆找平层
300厚钢筋混凝土层
混凝土垫层
素土夯实

Down
Down
Up
Up
Up
Down
Up
Down
Up
Down
Down
Up

刂用不同高度的特点匹配不同的功能，将居民生活、游客观光、文化展览在此地交汇共生。

rovide different functions by making the best of the characteristic of different heights, nd realize convergence and symbiosis of life of residents, tourism and cultural xhibition in this place.

传统乡村聚落复兴研究

STUDY ON REVITALIZATION OF TRADITIONAL RURAL SETTLEMENT

张雷 王铠

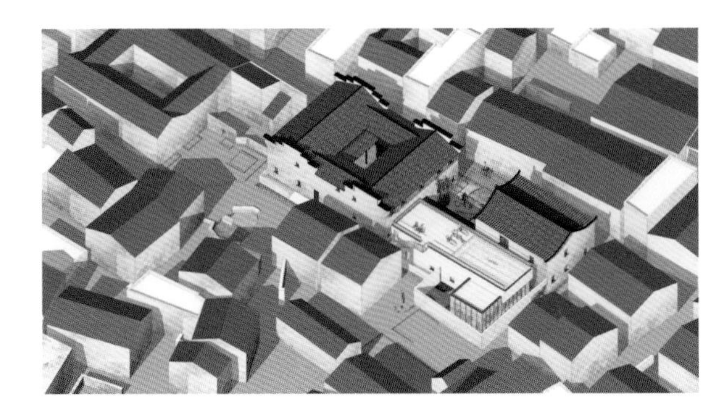

1. 课程说明

课程从“环境”“空间”“场所”与“建造”等基本的建筑问题出发，对乡村聚落肌理、建筑类型及其生活方式进行分析研究，通过功能置换后的空间再利用，从建筑与基地、空间与活动、材料与实施等关系入手，强化设计问题的分析，强调准确的专业性表达。课程通过设计训练，让学生达到对地域文化以及建筑设计过程与方法的基本认识与理解。

2. 基本建筑

作为整个西方现代文明的重要组成部分——对新技术和新材料的应用，程式化的功能考虑，结构和界面分离的理性思考……现代建筑的诞生和发展是对建筑古典传统字面意义上的反叛与解放。现代建筑打破戒律，消除等级，进行技术革命……使得属于少数人的纪念碑不那么重要了，而“大众”成为主要的“服务”对象。现代建筑和其他人造物一样，是满足大众需求的工业化产品。

现代主义强调建立秩序的重要性，重视事物内在规律的探索和重构，试图以整体、统一、协调的视觉秩序来积极地描述我们所生活的世界，并以简洁的几何关系予以表达。当代中国建筑师所面临的“速度”和“数量”的压力是空前的，这让我们自然地去求助于现代建筑的基本问题：用最合理、最直接的空间组织和建造方式解决问题，以普通的材料和通用的方法去回应复杂的适用要求，从普通的素材中发掘具有表现力的组织关系，这些正是设计所应该关注的基本原则。而这更应该成为今天中国社会快速成长中大规模建设要求的适用的工作策略，也有利于人类有限资源的利用和配置。

课程教学立足“环境”“空间”“场所”与“建造”等基本的建筑问题，强调作为建筑设计核心的基本甚至简单逻辑法则，去推演复杂的现象或设计问题。

3. 乡土聚落

“地域性”这个被当代建筑师反复提及和热衷的概念变得越来越宽泛，在大多数情况下，“地域性”仅仅是装饰蛋糕的水果花边。标签化的地域符号、博物馆式的消极保护、旅游经济的粗糙产品包装，不断地消磨着地域建筑文化的内在整体活力。

中国城市之前 30 年的大发展主要成就在城市，而真正属于我们自己的建筑文化传统在于以农业文明为根基的乡土聚落、有机整体的城乡关系以及持续数千年的发展传承模式，即一种处于“文化自觉”的地域性。反思现代、回归传统，是当下中国乡建活动众多实践方向的共同价值纽带。无论出于何种初衷，到乡下做事，无法避免遭遇中国乡村的普遍现实：落后的经济技术环境、依然牢固的宗族和地缘社会关系，以及整体又独特的乡土自然与人文环境。

现实的人的本性需求，是差异、交流、延续的地域性活体的内在动因。课程期望设计组成员能够以生活的参与者、体验者的姿态，去学习和领悟那些困境中蕴含的时间下凝聚的群体智慧，以及生活中的快乐与舒适、辛劳与节俭、感动与自豪、记忆与希望……

4. 向没有建筑师的建筑学习

基于现代城市工业化、资本运行逻辑导致的专业化标准答案，对于“地域性”往往失效，乃至有害。“时间”和无名的建造者给了我们地域建造的答案，只不过受当代工具文明的冲击，这些答案表述的语境变得含糊不清，难以辨识。面对那些看似粗陋的构造、衰败的房屋、木讷的工匠，我们在出于礼貌的同情和怜悯之后提议的“创新”和“设计”能够经得住时间的考验吗？答案往往是否定的，明智的选择是拂去厚厚的尘土，去学习时间沉积的智慧，去延续地域传统的当代语境。

课程设计分为组织现场调研和设计 / 研讨两部分。集中一周的现场调研一方面要求同学们用专业的方式去观察、记录并重绘改造对象的物质空间形式，分析其材料建造逻辑；另一方面结合地方民宿产业的调查，感受物质空间与有关日常生活经验的相关性，进而提议真正属于乡土聚落的改造利用方式。

仔细地观察乡村的真实面貌，通过设计“努力强化那些看上去有价值的，改掉那些不舒服的，重新创造那些我们觉得快要失去的东西”。这是一个不断学习的过程：向没有建筑师的建筑学习，向乡土生活学习。课程期望建立的这种“学习”的态度，成为整个设计 / 研讨的主线。

5. 时间 / 空间

作为工程技术角色的建筑师，必须将乡村聚落复兴的社会学思考和现实制约作为工作的起点，从一栋房子、一个院子开始，去了解和思考乡村的孰优孰劣、孰是孰非，去了解和思考乡村真正需要的是什么。任何地区打动人心的建成环境特征，都是时间沉积的力量，时间是生活的熔炉，是群体智慧的发酵池。相对于无法抗拒的时间，建筑师所精心构造的“空间”设计，往往更加个人化，而无法面对生活的考验。在后期的设计 / 研讨阶段，课程设计的空间创造被严格地限制，转而将注意力集中在处理“新”与“旧”的辩证关系——“新”形式是“旧”重新建立的空间边界，将过去的时间重新引向现实。

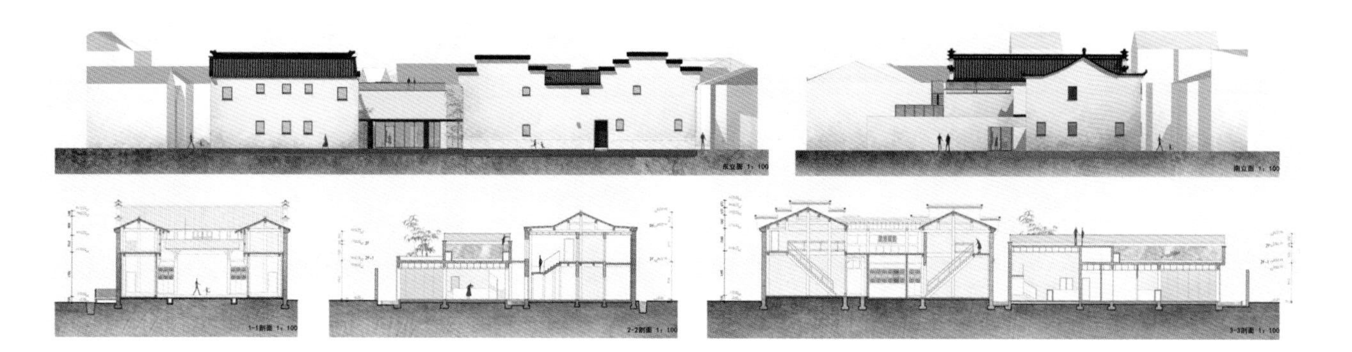

1. **Course Description**

The course starts from fundamental architectural issues such as "environment", "space", "site" and "construction", conducts analysis and research on texture, building type and life style of rural settlements, and aims to strengthen analysis on design problems and emphasize accurate professional presentation on basis of relations between building and base, space and activity, and material and execution, through reutilization of space after function replacement. The couse made the students obtain basic cognition and understanding on regional culture as well as architectural design process and methodology through design training.

2. Basic Architecture

As an important part of entire modern civilization in western countries – application of new technologies and new materials, consideration on stylized functions, and rational thinking on separation between structure and interface··· the birth and development of modern architecture are rebellion and liberation to classical traditional literal meaning of architecture. Because modern architecture broke the commandments , eliminated the hierarchy, and carried out technological revolution···, so the monument that belongs to a few people is no longer so important, and "the public" has become the primary subject to be served. Similar to other man-made articles, mordern buildings are industrialized products that are made to meet demands of the public.

The modernism emphasizes the importance of establishing sequence, pays attention to exploration and reconstitution of inherent rules of objects, and attempts to actively describe the world where we live with integrated, uniform, and coordinated visual sequence, and to present it with simple geometrical relations. The pressure in terms of "speed" and "quantity" faced by modern Chinese architects is unprecedented, and it naturally makes us resort to basic questions of modern architecture: solve problems with most rational and most direct ways of space organization and construction, respond to complex application requirements with normal materials and common ways, explore organizational relations of largest expressive force from normal materials, and these are basic principles that should be focused on in design work. And furthermore, it should become the work strategy suitable for the demand of large-scale construction during rapid growth of the Chinese society today, and it is also conducive to utilization and allocation of the limited resources of mankind.

Based on basic architectural issues of "environment", "space", "site", and "construction", the course teaching emphasizes fundamental and even simple logic rules that are core of architectural design to deduce complex phenomena or design issues.

3. Rural Settlements

The concept "regionalism" that is referred to repeatedly and loved by modern architects become wider and wider, and in most cases, the "regionalism" is just the fruit lace decorating the cake. Labeled regional symbols, museum-like passive protection, and coarse product package of tourist economy are constantly wearing down the overall inherent vitality of regional architectural culture.

Main achievements during the past three decades in China are made mainly in cities, while our real cultural tradition of architecture lies in the rural settlements being rooted in agricultural culture, the organic integral of urban and rural relations, as well as the development and inheriting mode that has been lasted for thousands of years, which is a type of regionalism in "cultural consciousness". Reflecting on modern times and getting back to the past are the common value link of various practice directions for current rural construction activities in China. No matter what original intentions are based on, working in rural areas will inevitably be confronted with the common reality in rural areas of China: lagging-behind economic and technical environment, clan, regional, and social relations that are still strong, as well as integrated and unique rural nature and cultural environment.

Inherent demand of realistic human being is the internal driving force for differentiated, interacting, and continued regional living creatures. The course expects design team members, in posture of participants and experiencers of life, to learn and comprehend the wisdom of crowds that has been condensed over time in those hardships, and the happiness and comfort, toil and frugality, touching and pride,

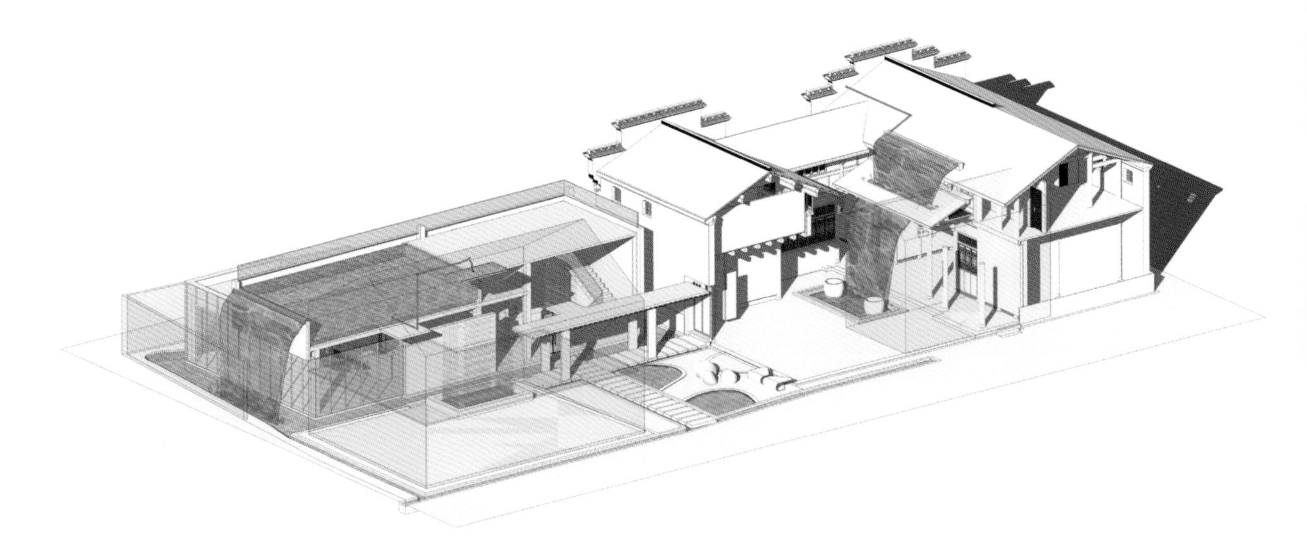

memory and hope··· in daily life.

4. Learning from Buildings Not Designed by Architects

The professional standard answers that are produced on basis of modern urban industrialization and capital operation logic are often invalid, even detrimental to "regionalism". The "time" and anonymous builders give us answer about regional construction, only that impact of modern tools on civilization has made context for presenting these answers become obscure and difficult to be identified. In the face of those seemingly rough construction, dilapidated houses, and dull craftsman, can the "innovation" and "design" proposed by us after politeness-based sympathy and compassion withstand the test of time? The answer often is negative, and the wise choice is to flip away the thick dust, to learn the wisdom deposited over time, and continue the modern context of regional traditions.

The course design consists of two parts of field investigation and design/discussion. The one-week field investigation, on the one hand, requires students to observe, record, and re-draw the form of physical space of the object to be reconstructed, and to analyze construction logic of materials; and on the other hand, by combining with the investigation on local homestay industry, to feel the relations between physical space and daily life experience, and come up with a way of reconstruction and utilization that really belongs to rural settlements.

Carefully observe real features of rural areas, and strive to enhance those seem valuable, modify those look comfortable, and recreate those things that we think they are going to disappear through design. This is a constant learning process: learn from buildings not designed by architect, and learn from rural life. The course is intended to establish this "learning" attitude, and make it the main line of entire design/discussion.

5. Time/Space

An architect undertaking the role of engineering technology must make sociological thinking and restriction of reality about revitalization of rural settlements as the starting point of his/her work, and understand and ponder over rural areas for strengths and weakness, what is right or wrong, and what do rural areas really need, by starting from one building, and one courtyard. The touching building environment feature in any areas is the force deposited over time, and time is the melting furnace of life, the fermenting tank for wisdom of crowds. In comparison with the overpowering time, the "space" design configured by an architect deliberately is often more personal, and cannot face with the experience of life. During later phases of design/discussion, space creation in the course design will be restricted strictly, and attention will be focused on the handling of the dialectical relationship between "old" and "new" – the "new" form is the spatial boundary recreated by the "old", and it redirects the past time to reality.

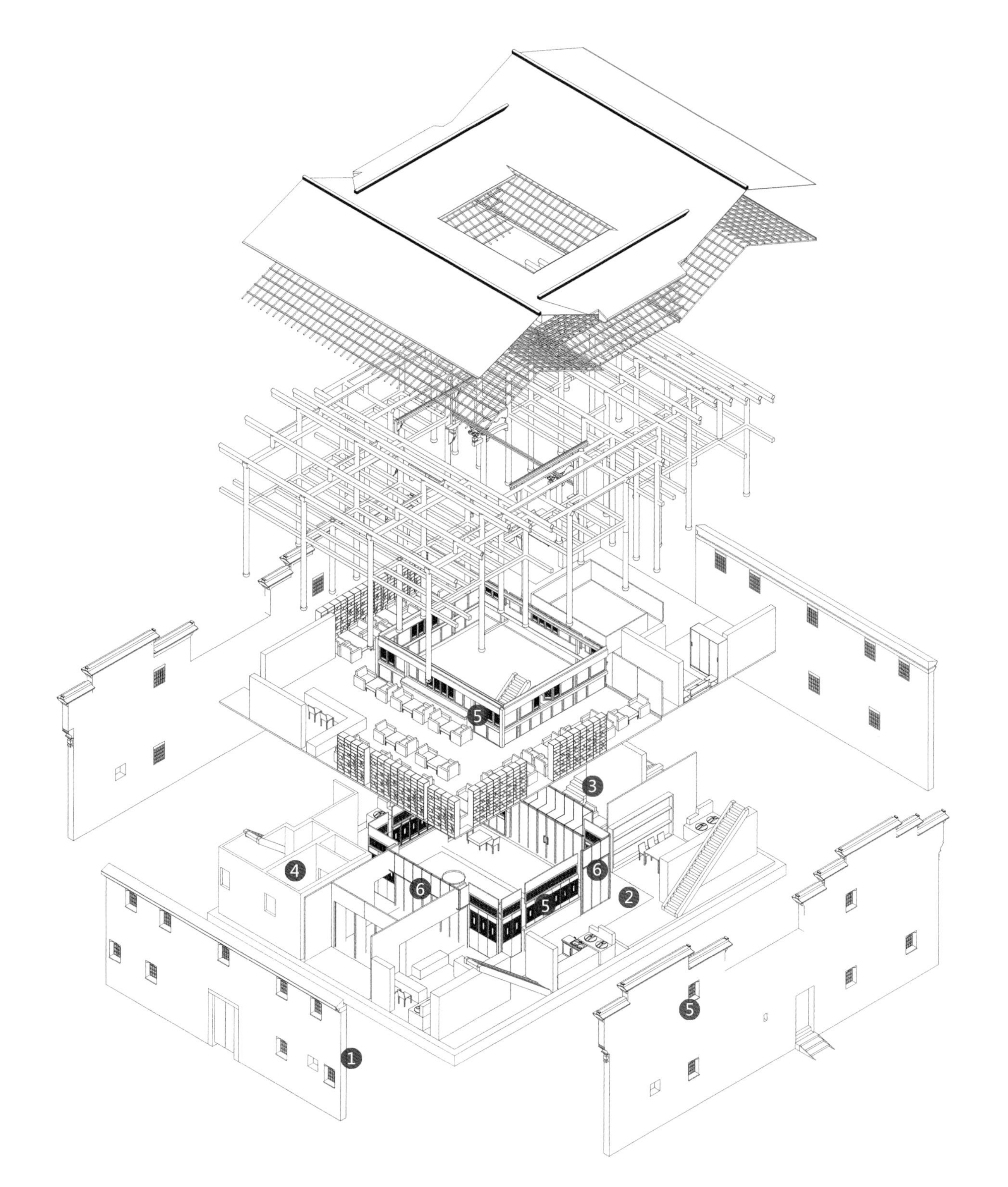
5
3
4
6
6
2
5
5
1

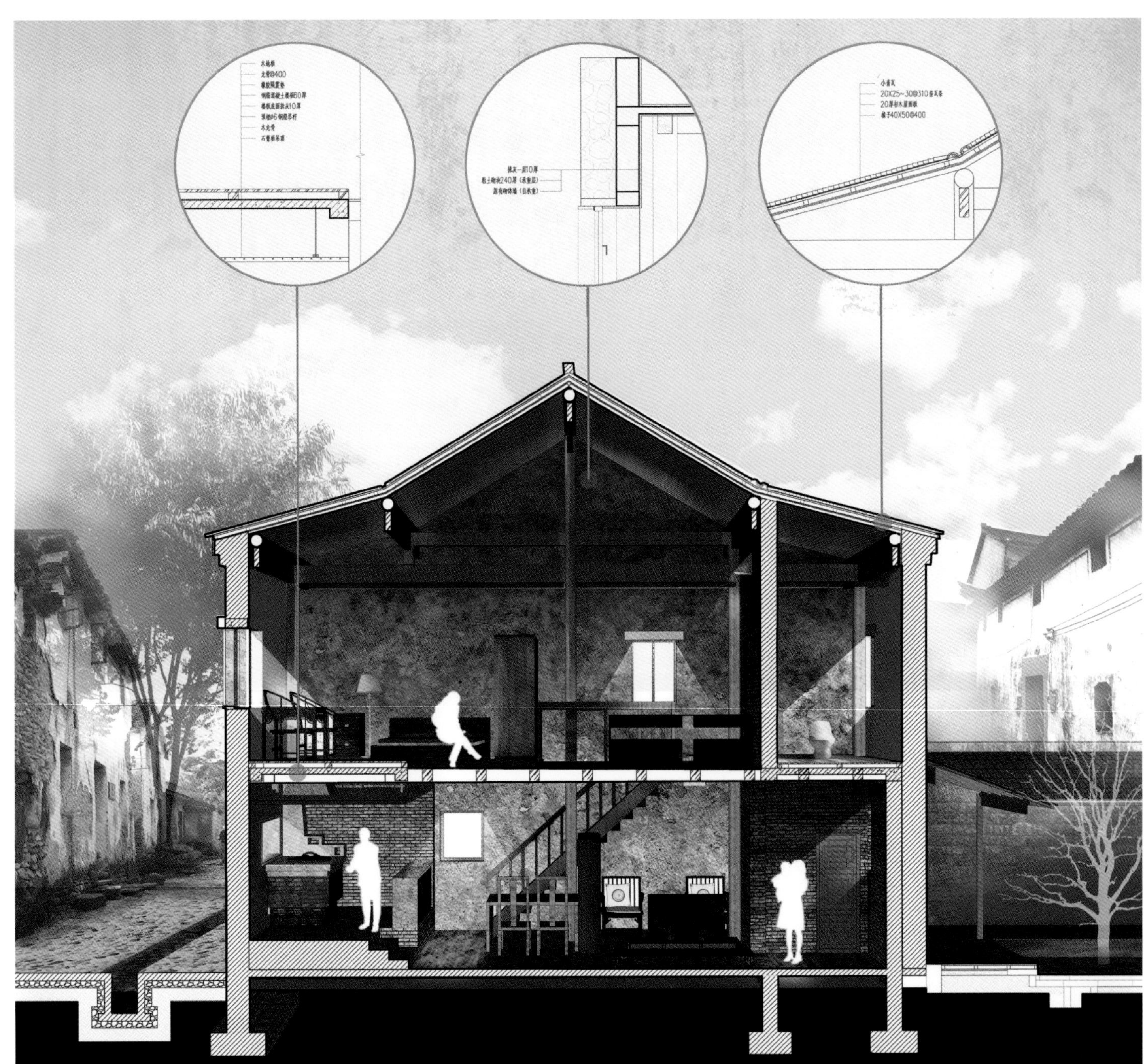

结合地方民宿产业的调查，根据物质空间与有关日常生活经验的相关性，围绕原有建筑内的土灶组织游客所需的住宿、餐饮、体验等功能，将传统合院式民居改建成8套亲子民宿。

In combination with investigation on local homestay industry, based on the correlation between material space and relevant daily experience, the design organizes the accommodation, catering and experience and other functions required by the tourists based on the cooking hearths in the original building, and has transformed traditional courtyard-style houses into 8 parent-child homestay facilities.

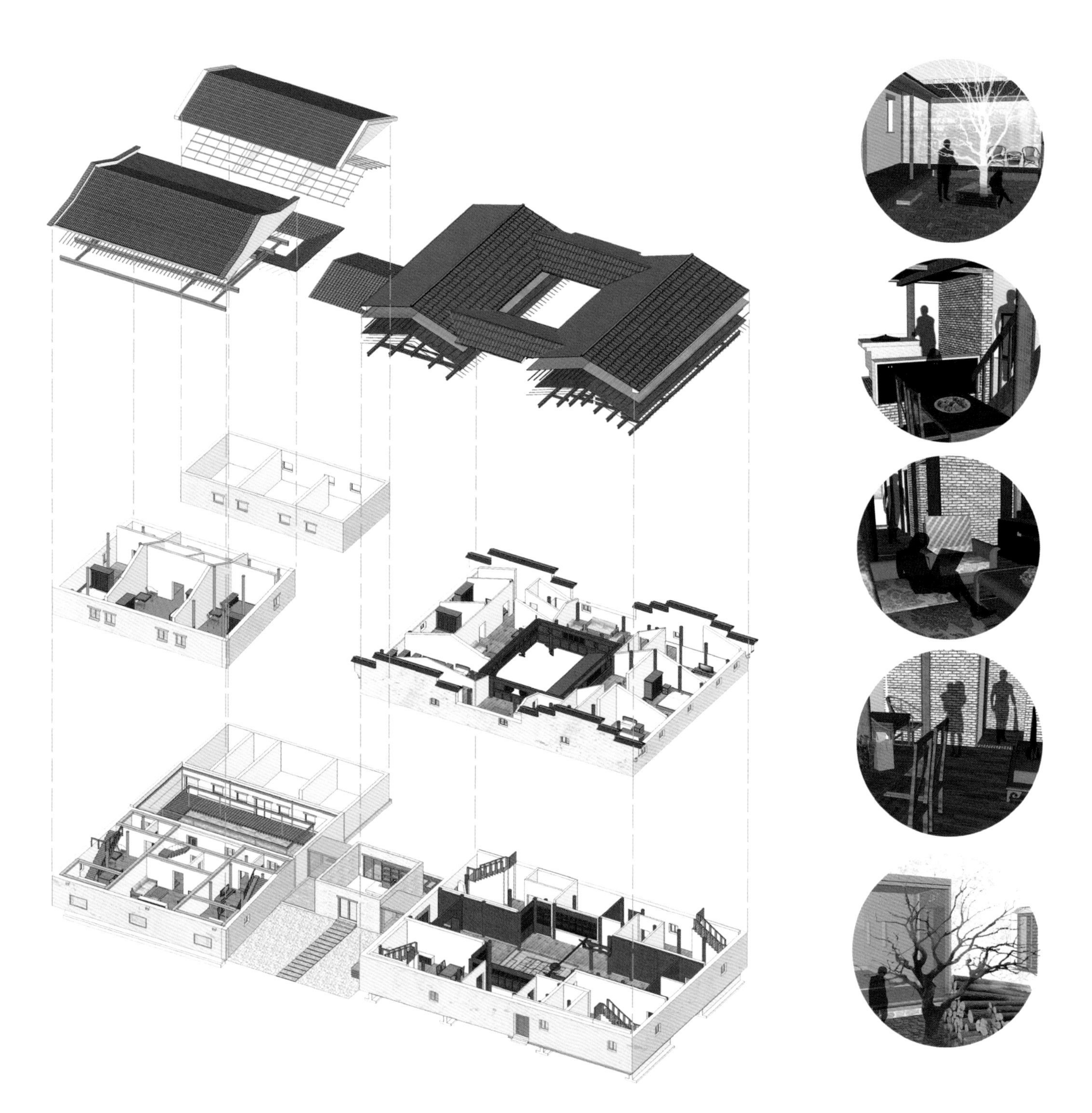

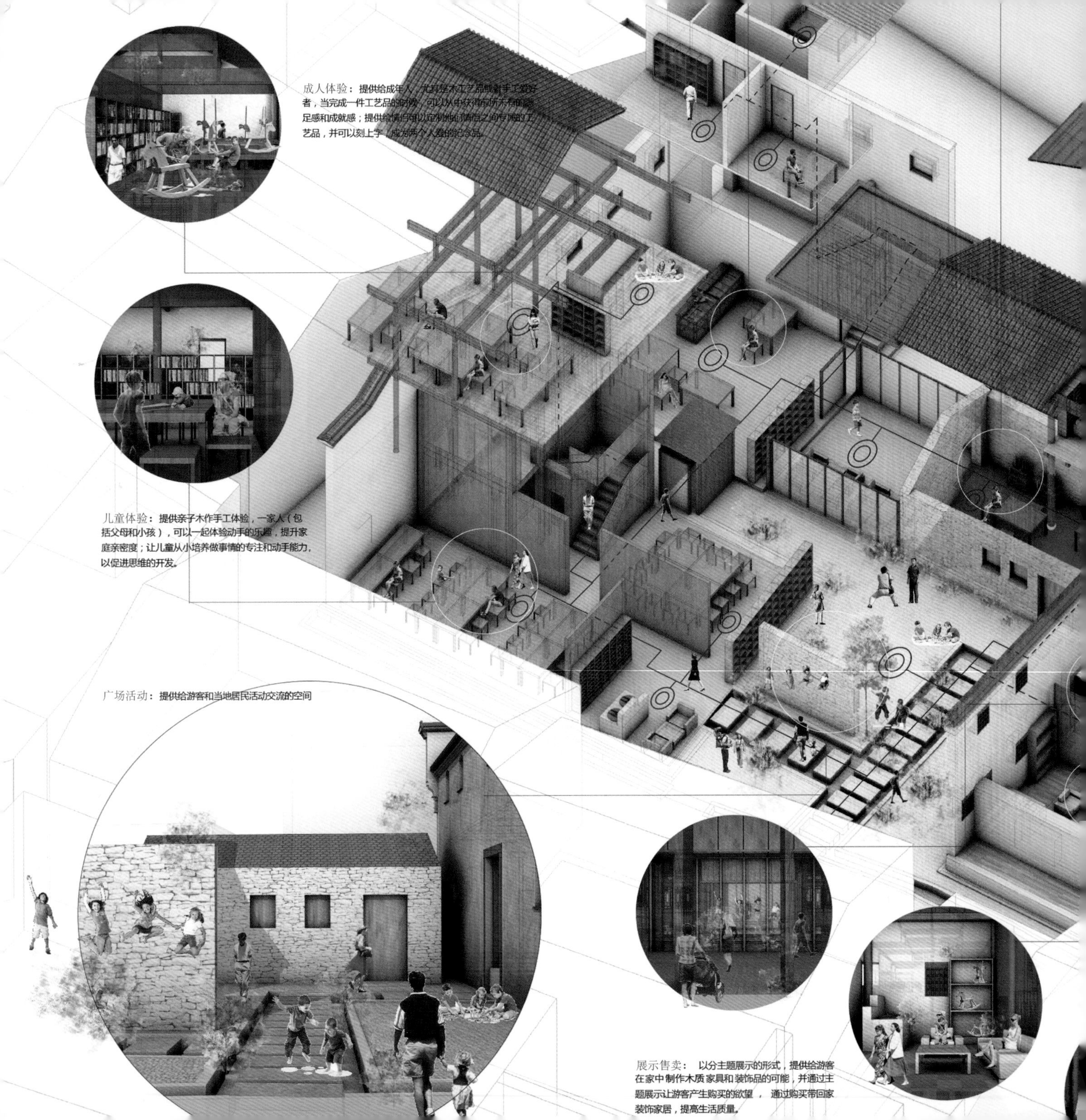
成人体验：提供给成年人，尤其是木工艺品或者手工爱好者，当完成一件工艺品的时候，可以从中获得前所未有的满足感和成就感；提供给情侣可以定制他们情侣之间专属的工艺品，并可以刻上字，成为两个人爱的纪念品。
儿童体验：提供亲子木作手工体验，一家人（包括父母和小孩），可以一起体验动手的乐趣，提升家庭亲密度；让儿童从小培养做事情的专注和动手能力，以促进思维的开发。
广场活动：提供给游客和当地居民活动交流的空间
展示售卖：以分主题展示的形式，提供给游客在家中制作木质家具和装饰品的可能，并通过主题展示让游客产生购买的欲望，通过购买带回家装饰家居，提高生活质量。

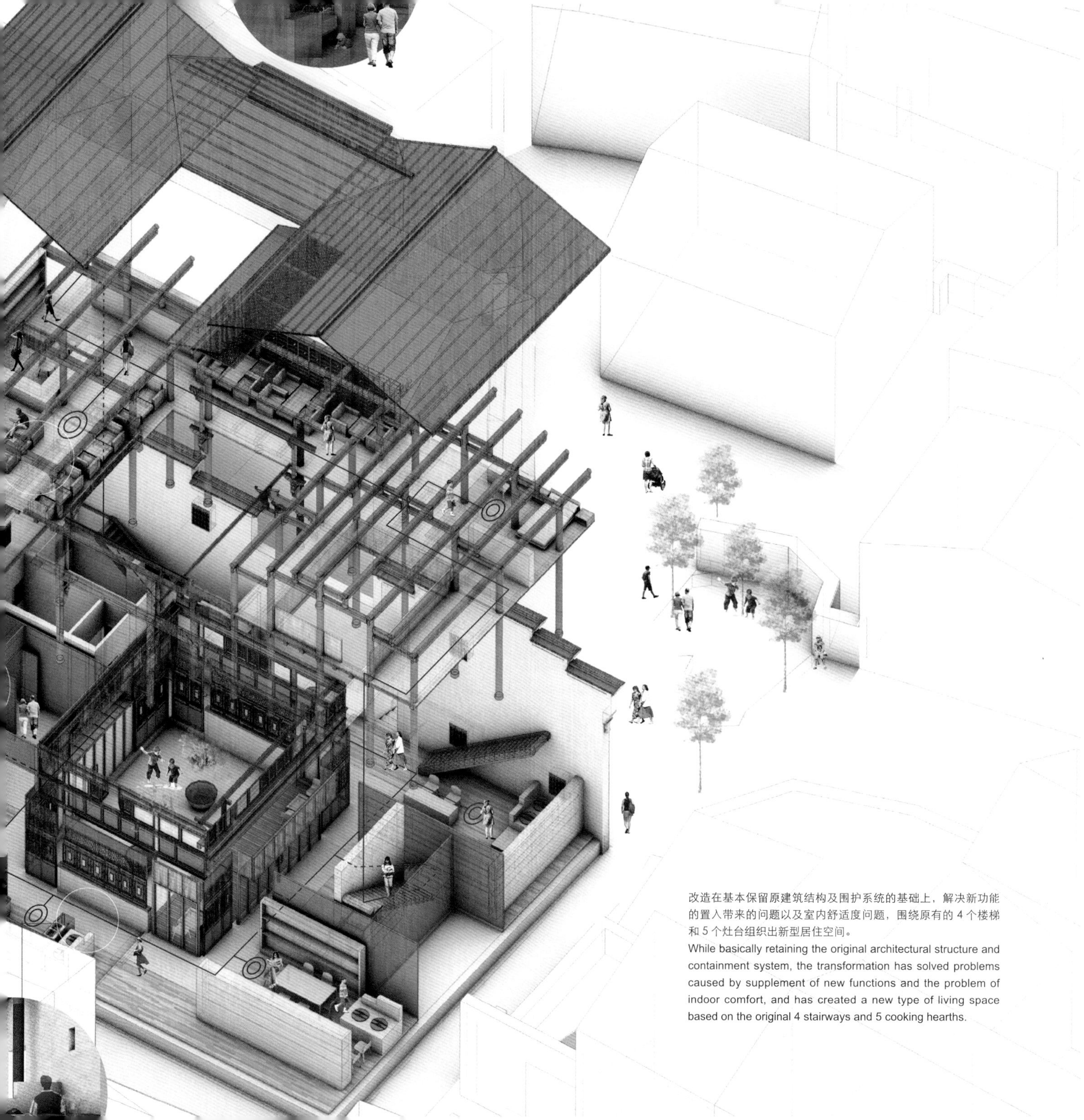

改造在基本保留原建筑结构及围护系统的基础上，解决新功能的置入带来的问题以及室内舒适度问题，围绕原有的 4 个楼梯和 5 个灶台组织出新型居住空间。

While basically retaining the original architectural structure and containment system, the transformation has solved problems caused by supplement of new functions and the problem of indoor comfort, and has created a new type of living space based on the original 4 stairways and 5 cooking hearths.

南京大学体育中心设计

DESIGN OF SPORTS CENTER OF NANJING UNIVERSITY

郭屹民

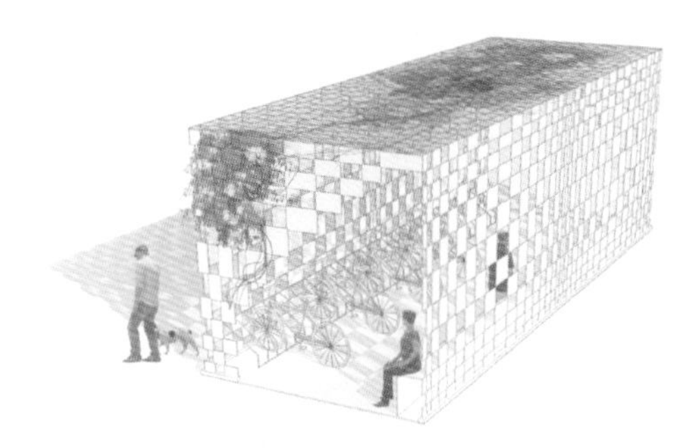

1. 教学内容

北苑校区现有体育馆设施陈旧，功能单一，已无法满足现有师生更加丰富多彩的健身锻炼需求及社交活动要求。现拟在原体育馆建筑用地范围内改扩建体育中心。

2. 教学目标

（1）掌握结构设计基础知识，并会进行结构分析和结构设计。

（2）了解结构的材料与建造，并会通过材料和建造进行建筑构造设计。

（3）了解结构设计与功能、场地的关系，并会进行与功能相关的建筑设计。

3. 教学总结

结构设计课对建筑空间的设计是在学生初步掌握结构形态设计方法的“跨度形态设计”之后展开的。与只需要对应力学约束下的形态设计的结构形态不同的是，建筑空间需要面对包括来自周边环境、功能、建造、经济、文化、社会等诸方面的影响。结构力学作为影响建筑空间要素之一，必须在与其他要素的协同中对既有纯粹力学范畴上的合理性进行必要的调整与变化，才能最终使建筑空间在各要素之间获得平衡，来实现建筑空间上的合理性。学生能够从结构形态到建筑形态的设计过程理解两者之间的共性与区别。

考虑到建筑空间形态设计的周期相对较短，以及我们希望尽可能地将前一阶段的结构形态、节点设计、建造等内容体现在对空间的表达上，因此，设计任务被限定在南大校园中学生所熟知的原有体育中心的改建上。这样有利于缩短调研和基地踏勘的时间，从而将更多的时间用于分析与研究上。学生需要反复通过对结构形态与空间和基地环境、边界条件、使用功能、施工条件作出反馈和调整，来最终确立建筑的内外空间。在这种与建筑要素相对应的反复过程中，学生会意识到结构形态也在逐步向建筑形态迈进。

在教学过程中，我们坚持要求学生以模型的方式来研究整体与局部的关系，这样也有利于随之发现设计过程中结构设计的正确与否。体育中心的大小空间的组合，本身就给结构设置了不小的障碍，但同时也给全新的结构与空间的发生提供了极大的可能。尽管是 2 人一组，但 8 周时间的结构设计课无论在时间上还是在领悟与设计上，对于学生而言都非常具有挑战性。我们在教学指导中尽量避免趋同而鼓励多样化的思路拓展，激励学生以创新的姿态面对全新的设计过程。

事实上，以建筑师的角色体验结构与空间造型的构成，本身就是一个理性与感性重新磨合的过程。没有复杂的公式，有的是结构对造型的再创造。通过这样的课题，学生能够真正理解建筑造型并非是随心所以、为所欲为的个人化行为。在他们所熟知的场地、功能之外，还有技术层面的约束。这种体会也将使得学生有机会通过结构这一视角反思建筑空间的本质，从而触发他们对建筑更加深入的思考。

1. Teaching Content

There has an existing gymnasium with outdated facilities and unitary function at Beiyuan campus. It cannot meet the demand of faculty and students on more colorful fitness exercises and social activities. Now it plans to improve and expand the sport center within the land area of the existing gymnasium.

2. Training Objective

(1) To grasp the basic knowledge of structural design, and to be able to conduct analysis on and design of structure.

(2) To understand the materials and construction of structure, and to be able to conduct building structure design with the materials and through construction.

(3) To understand relations among structural design, functions and the field, and to be able to conduct function-related architectural design.

3. Course Conclusion

The design of architectural space for structural design course is carried out after students have preliminarily grasped the design method of structural form of “span form design”. Different from structural form of form design that only corresponds to mechanical constraints, architectural space shall face with the influences on habitation coming from such aspects as ambient environment, function, construction, economy, culture, and society. As one of those elements affecting architectural space, structural mechanics must make necessary adjustment and change to rationality of existing pure mechanical considerations in collaboration with other elements, so as to obtain balance of architectural space among various elements, and to achieve rationality of architectural space. And students may understand the commonality and difference between structural form and architectural form in the process of design.

Given that the form design of architectural space lasts only for a short period, and that we want to reflect matters taught in previous period including structural forms, detail design, and construction through expression of space to the greatest extent possible, the design task is limited to the improvement work of an existing sports center within the campus of Nanjing University that is familiar to students. In this way, we can reduce the time required for investigation and field survey, and spend more time on analysis and research. Students are required to feedback and adjust structural form and space and field environment, perimeter conditions, functions,

and construction conditions repeatedly, so as to finally determine interior and exterior spaces of the architectural space building. In this repeated process of responding to architectural elements, students will realize that structural form is also stepping closer to architectural form gradually.

In the process of teaching, we persistently require students to study relations between entirety and local parts with models, which is also conducive for them to determine whether the structural design is right or not in the process of design. Combination of spaces of different sizes of the sports center itself places a number of obstacles to structure, but it also provides infinite possibility for occurrence of new structures and spaces. Although each group consists of 2 students, 8-week structural design course means lots of challenges for students in terms of time, as well as comprehension and design. During teaching instruction, we try the best to avoid similarity, and encourage them to seek for diversified ideas, and motivate students to face with the brand-new design process in posture of innovation.

As a matter of fact, experiencing the formation of structural and spatial modeling in the role of architect is a process of refining rationality and sensibility. Without complicated formulas, there is only the re-creation of modeling with structure. Through this course, students can really understand that architectural modeling is by no means personal behavior that we can do what we want. There are restraints in term of technology in addition to that of field and functions that we are familiar with. Such experience also allows students to have the opportunity to reflect on the nature of architectural space in the view of structure, thus trigger more in-depth thinking on architecture by them.

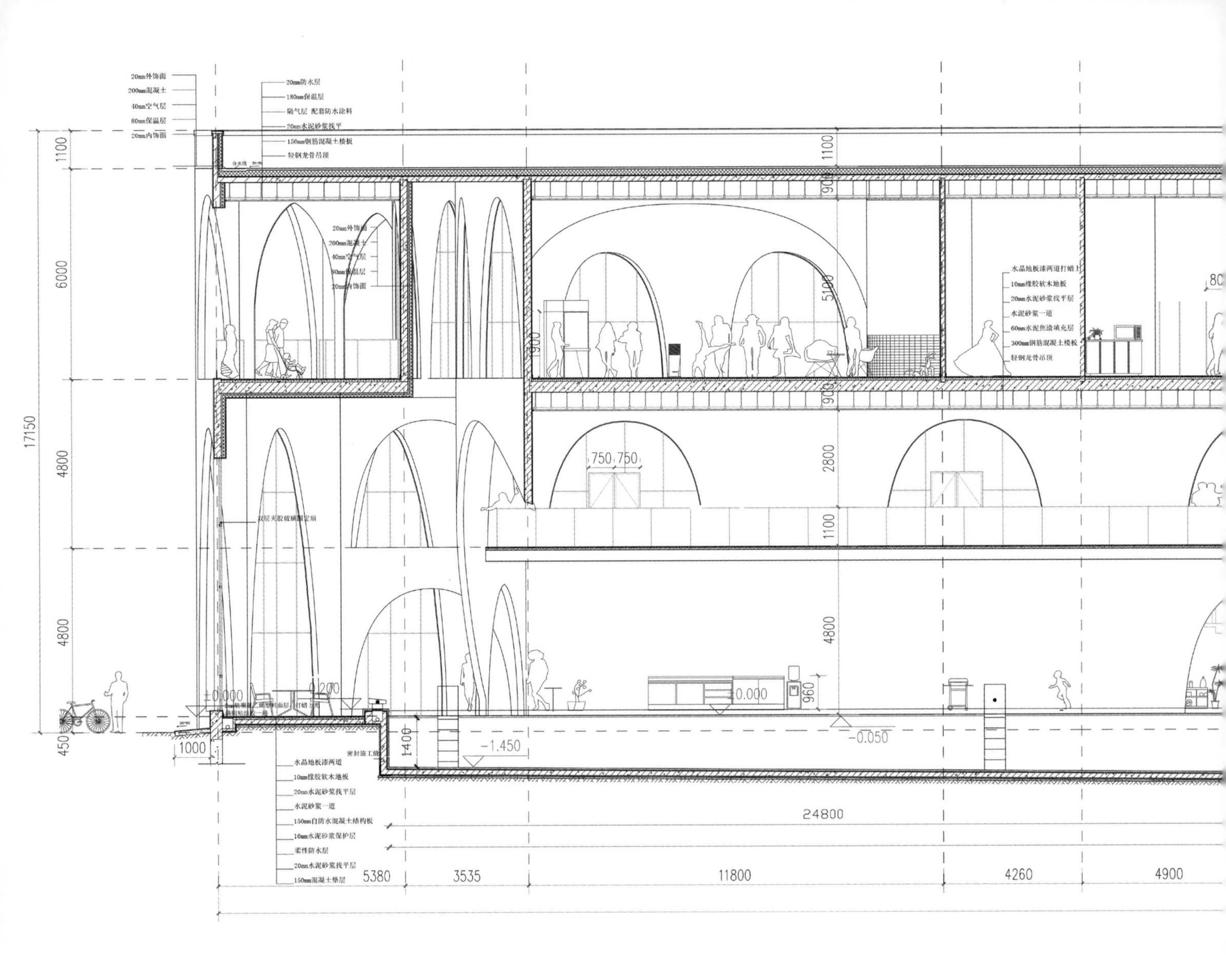

20mm外饰面
200mm混凝土
40mm空气层
80mm保温层
20mm内饰面
20mm防水层
180mm保温层
隔气层 配套防水涂料
20mm水泥砂浆找平
150mm钢筋混凝土楼板
轻钢龙骨吊顶
水晶地板漆两道打蜡
10mm橡胶软木地板
20mm水泥砂浆找平层
水泥砂浆一道
60mm水泥焦渣填充层
300mm钢筋混凝土楼板
轻钢龙骨吊顶
双层夹胶玻璃固定扇
密封施工缝
水晶地板漆两道
10mm橡胶软木地板
20mm水泥砂浆找平层
水泥砂浆一道
150mm自防水混凝土结构板
16mm水泥砂浆保护层
柔性防水层
20mm水泥砂浆找平层
150mm混凝土垫层
17150
1100
6000
4800
4800
450
1000
5100
900
2800
1100
750
±0.000
0.200
-1.450
-0.050
1400
960
24800
5380
3535
11800
4260
4900

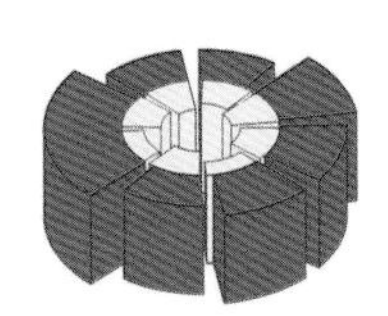
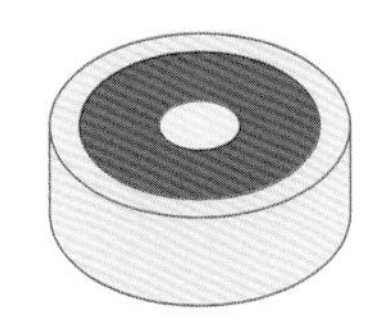

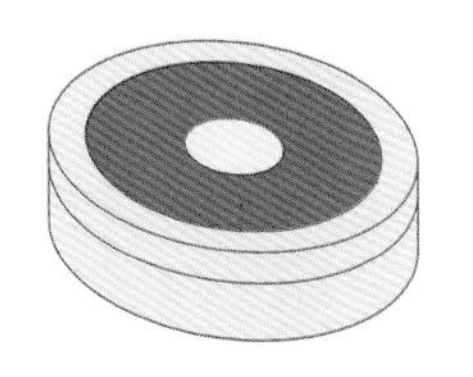
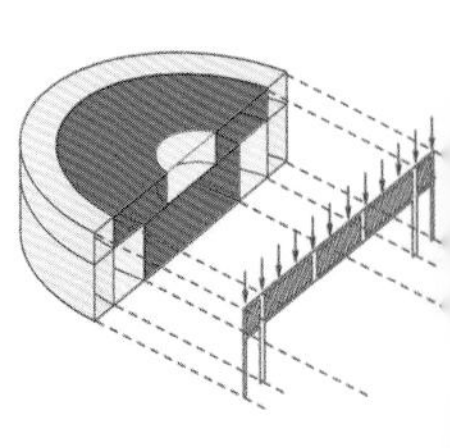

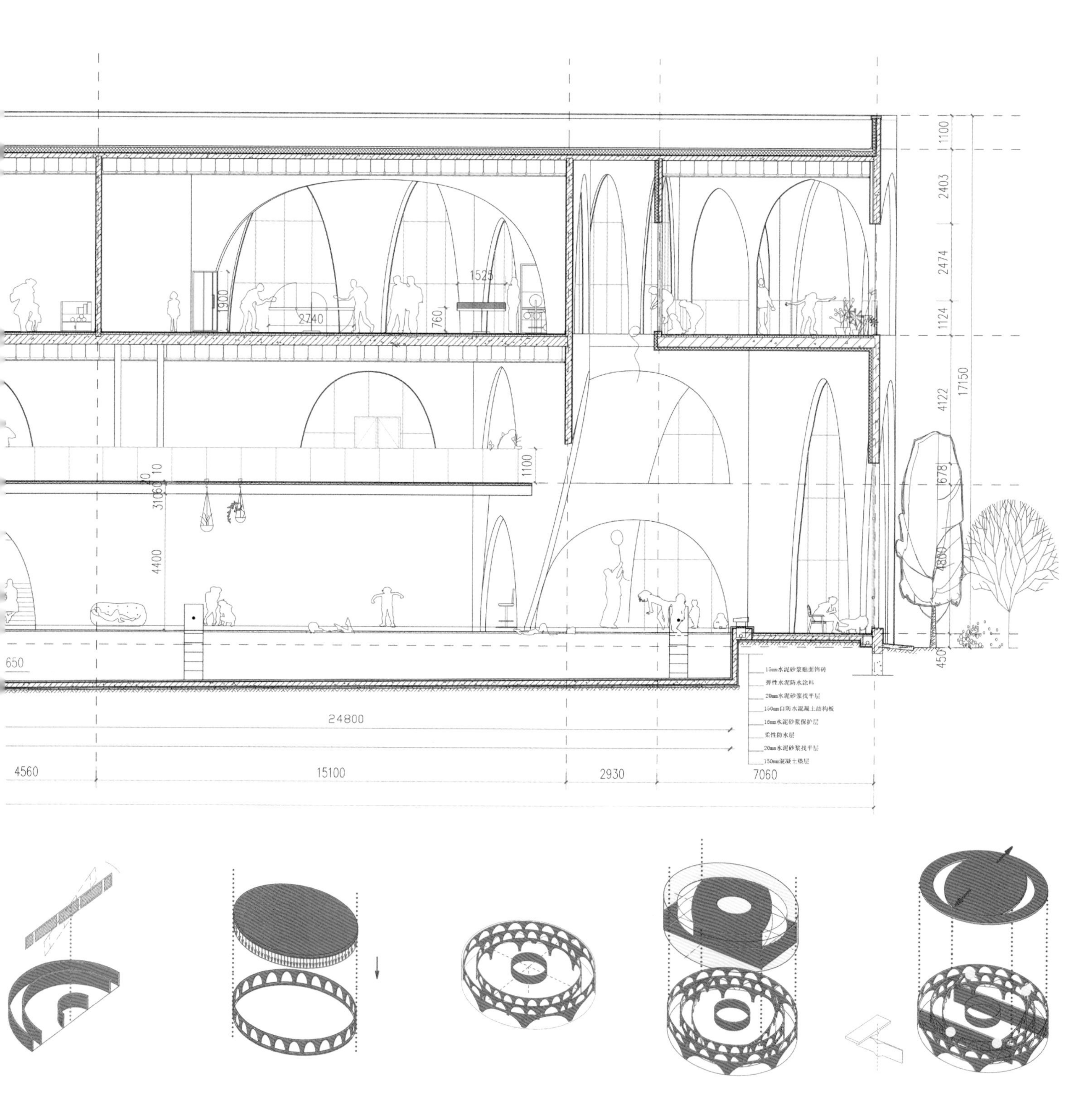
1100
2403
2474
1124
4122
17150
678
4800
450
1525
1900
2740
760
1100
3106
4400
650
15mm水泥砂浆贴面饰砖
弹性水泥防水涂料
20mm水泥砂浆找平层
150mm自防水混凝土结构板
16mm水泥砂浆保护层
柔性防水层
20mm水泥砂浆找平层
150mm混凝土垫层
24800
4560
15100
2930
7060

檐沟
轻型龙骨
防潮木板吊顶
钢筋混凝土梁
玻璃幕墙分隔龙骨
玻璃幕墙
150厚预制钢筋混凝土楼板
空气保温层
玻璃幕墙垫木

13.500
9.000
4.500

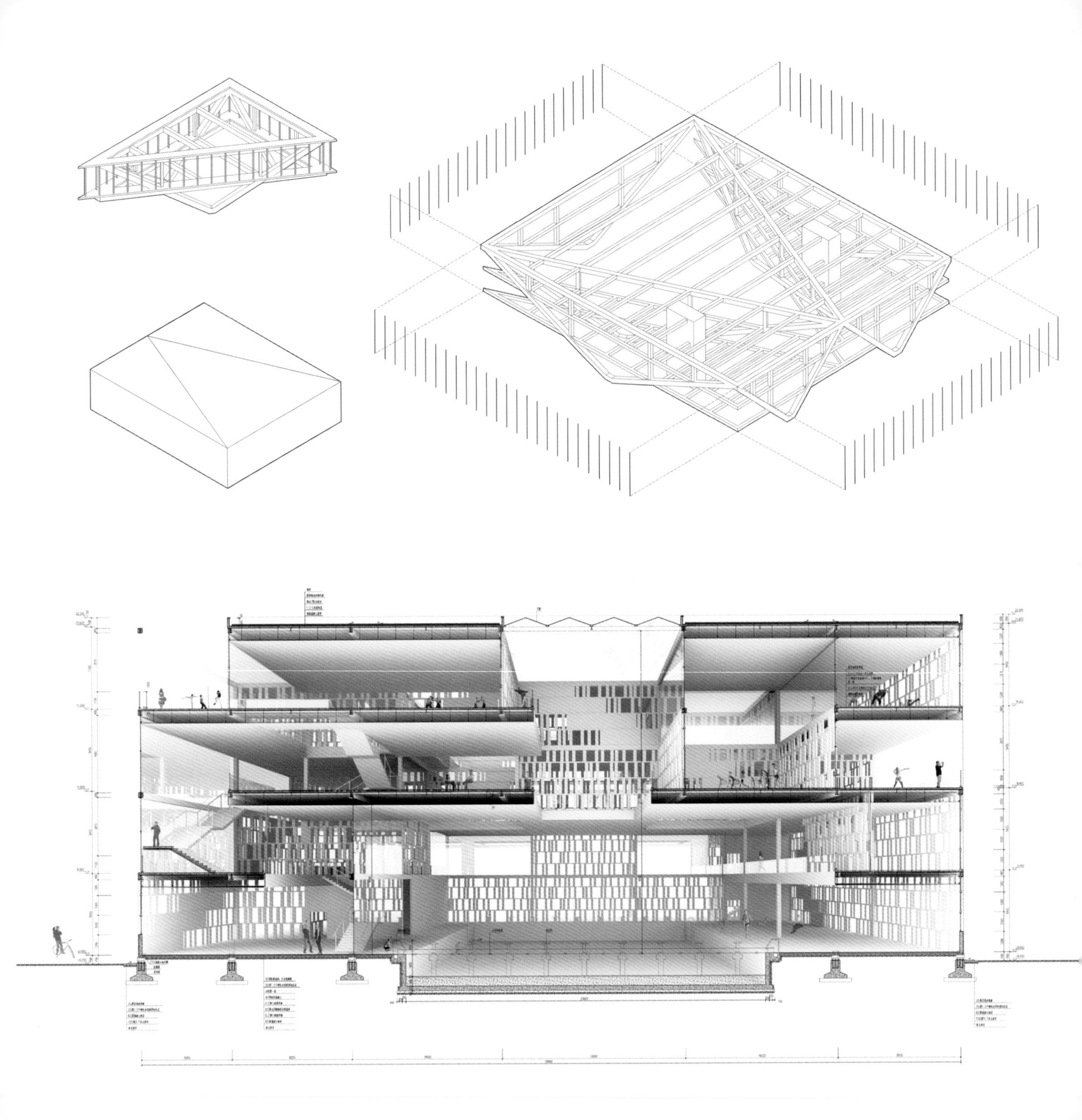

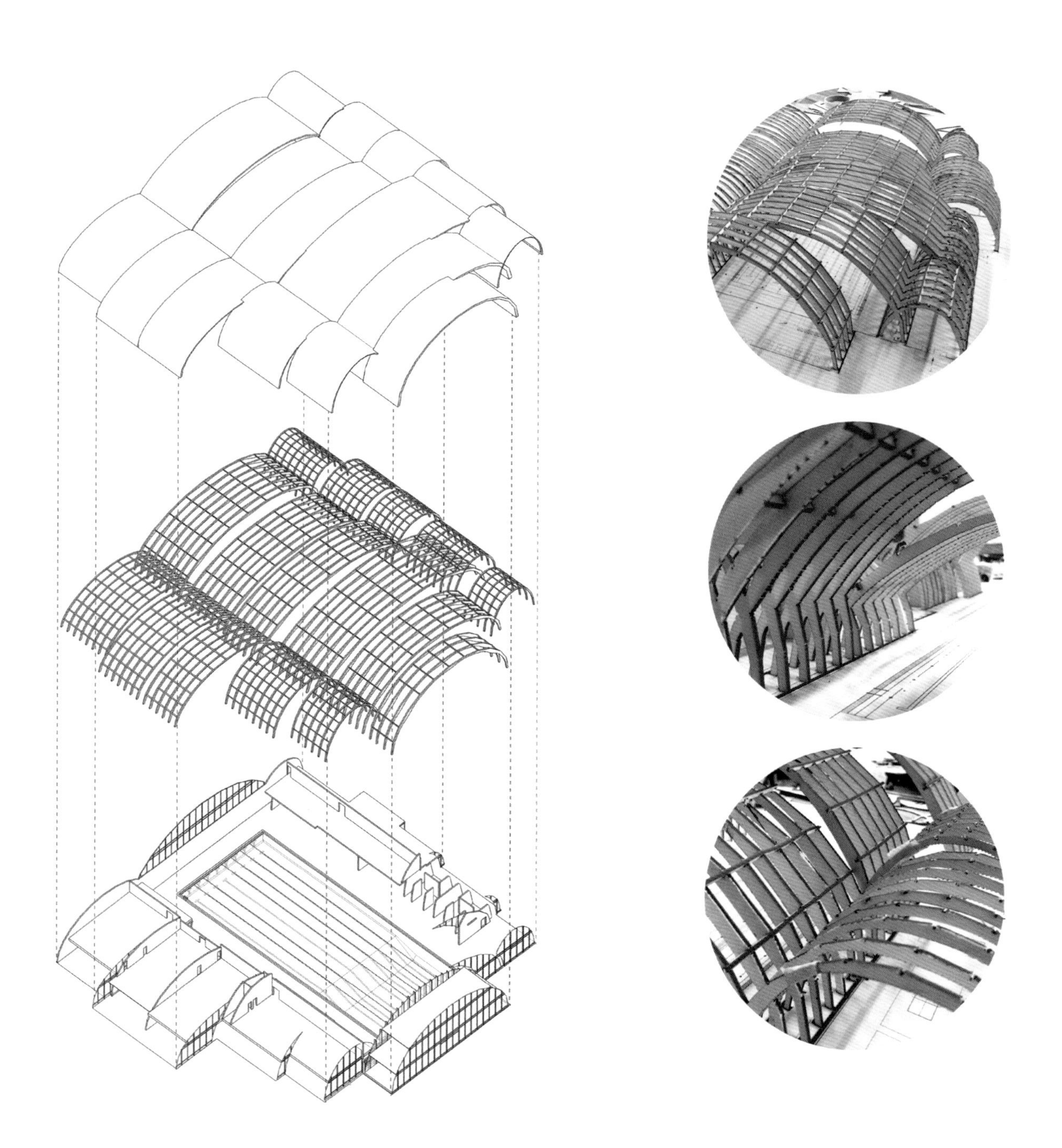

建构研究："低技建造"设计研究

DESIGN RESEARCH ON "LOW-TECH CONSTRUCTION" IN TECTONICS RESEARCH

傅筱　陈浩如

1. 建造技术研究课程回溯

南京大学建筑系研究生阶段的选修课程包括两个方向四门课程。一个方向是注重思维和设计方法训练的课程，包含"概念设计"和"城市设计"两门课程，另一个方向是注重基本逻辑和实际操作训练的课程，包含"基础设计"和"建造技术研究"两门课程。在十多年的教学中，建造技术研究由于受到资金、材料以及建造场地的制约，其间课程形式和内容变化较为明显，具体而言分为三个阶段。

2. 第一阶段（2000—2007）：木构建造

由于得到芬兰木业相关资助，在赵辰、冯金龙、周凌三位老师的带领下，学生开展了一系列的木构实地建造，期间的建造活动包含了蒙民伟楼地下室木构亭子建造、南京大学仙林校区步行桥建造、芬兰山地建造以及红山动物园系列木构小品建造等等。这阶段取得的课程研究成果是让学生通过亲身建造，真切体会到真实材料和真实尺度的意义，并建立起建造是衡量设计的核心标准的认知。

3. 第二阶段（2008—2014）："基础设计"的深化与发展

随着芬兰木业资助项目的结束，实际建造活动也暂告一段落。建造技术研究课程转为"纸上建筑"研究，课程以"基础设计"作业为基础，深入探讨两个问题，一是不同的结构类型对应的空间形态特征研究，一是设计概念与构造设计的关联性研究，课程由冯金龙、傅筱两位老师执教。2009 年，毕业于东京工业大学的郭屹民老师加入该课程，郭老师对结构颇有研究，因此开设了结构概念设计课程。至此，建造技术研究课程衍生为两门子课程，即建构设计和结构概念设计。建构设计主要是通过设计训练训练学生对设计概念与构造技术的关联性认识。结构概念设计主要在学习结构基本知识的基础上，让学生掌握结构与功能、结构与空间、结构与建造的关联关系。这一阶段的训练虽然是纸上谈兵，但是通过大量的技术图纸设计，让学生建立起了真正的设计图纸其实就是建造，每一根线条实际上就是建造的认知。

4. 第三阶段（2015—）：重回建造—竹构

2015 年，在周凌、赵辰两位老师的努力下，得到了相关资金和场地的支持，课程组得以重回建造。建造地点选择在浙江莫干山南路乡"60 亩农田服务设施规划"场地内，以竹结构为主实地建造"山野乐园"景观小品，供游客和儿童使用。这次建造活动有幸邀请到了有丰富竹构经验的"山上建筑工作室"主持建筑师陈浩如执教。在陈浩如、傅筱两位老师的带领下，学生完成了双亭、四面佛、三角亭、六角亭等竹构小品。这次建造活动仍然延续了"真实材料和真实尺度"的基本要求，让学生在设计和搭建过程中体验材料、尺度、空间、受力、建造的相互制约和促进作用。从教学全过程而言，学生体会最深的是构想一个纯粹的形式是容易的，但要整合出一个尺度合理、充分利用材料的力学特征同时易于实地建造的形式并非易事，而这些恰恰才是真正的形式。

一门课程如果能够较长时间地开设下去，说明课程探讨的一定是具有建筑学核心价值的问题，建造技术研究正是这样的课程，开设至今已然十五载。十五年中，课程的形式和内容随外在条件的改变而不断变化，可以预计这样的变化还将发生，我们也期待这样的发生，因为建造从来就是一种随条件改变而发生的活动！

1. Review on Construction Technology Research Course

Selective courses for graduate students at Architecture Department of Nanjing University include four courses in two directions. One direction covers courses emphasizing training on thinking and design method, including two courses of "Concept Design" and "Urban Design", and the other direction covers courses emphasizing training on basic logic and real-world practice, including two courses of "Basic Design" and "Construction Technology Research". During more than a decade of teaching practice, the form and content of the course of Construction Technology Research went through obvious change due to restriction of fund, materials, and construction site, and the process consists of three specific phases.

2. Phase I (2000–2007): Wooden Structure Construction

Thanks to the funding support from Finnforest, and under guidance and organization by three lecturers, namely Zhao Chen, Feng Jinlong, and Zhou Ling, students carried out a range of wooden-structure field construction activities, including construction of the wooden-structure pavilion at basement of Meng Mingwei Building, construction of footbridge at Xianlin Campus of Nanjing University, construction on mountainous land in Finland, and a series of wooden-structure landscape ornaments at Hongshan Zoo. Achievement of course research in this phase is allowing students to experience the meaning of real materials and real dimensions vividly, and to establish the awareness

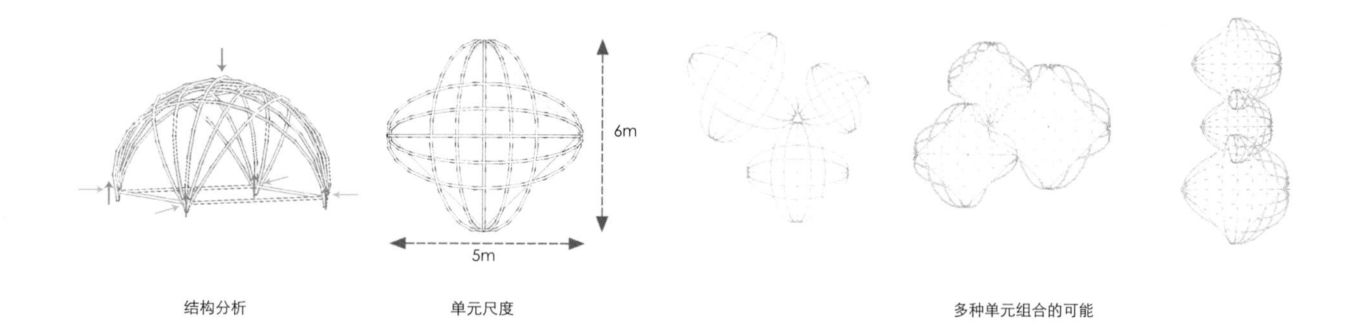

结构分析 单元尺度 多种单元组合的可能

that construction is a core criterion for design evaluation through in-person construction.

3. Phase II (2008–2014): Deepening and Development of "Basic Design"

The field construction activities came to an end as the funding project of Finnforest was finished. The course of Construction Technology Research was transformed into "paper construction" research; the course was based on "Basic Design" assignments, and explored two questions deeply. One is the research on morphological characteristics of space corresponding to different structure types, and the other one is the research on relevance between design concept and construction design. The course was taught by Feng Jinlong and Fu Xiao. In 2009, another lecturer Guo Yimin joined in this course; he was graduated from Tokyo Institute of Technology, and had deep insights on structure, so he opened the course of conceptual design of structure. Then, the Construction Technology Research course has evolved into two sub-courses, that is, Tectonic Design and Conceptual Design of Structure. Tectonic Design is intended to train students to understand relations between design concepts and construction technology through design training. Conceptual Design of Structure is intended to allow students to grasp relations between structure and function, structure and space, as well as structure and construction by learning basic knowledge about structure. Although trainings in this phase are mostly paper work, it enables students to establish the awareness through design of a large number of technical drawings that real design drawings are construction, and every line is also construction as a matter of fact.

4. Phase III (2015–): Back to Construction – Bamboo Structure

In 2015, thanks to efforts made by Zhou Ling and Zhao Chen, we received support of fund and field, so that the course team could get back to construction again. The construction site selected was within the planned field for "60-mu farmland service facilities" at Nanlu Town, South Moganshan Road, Zhejiang, and the project was building a "Wild Paradise" landscape ornament at the site with bamboo structure for tourists and children. We had the honor of having invited Chen Haoru, the principal architect from CITIARC Architecture and Design Office, to teach in this construction activity, who has abundant experience in bamboo structure. Under the guidance and organization of Chen Haoru and Fu Xiao, students completed bamboo-structure landscape ornaments including Twin Pavilion, Phra Phrom, Triangle Pavilion, and Hexagonal Pavilion. The construction activity still continued the basic requirement of "real materials and real dimensions", and allowed students to experience the mutual restriction and mutual promotion of materials, dimensions, space, force, and construction in the process of design and erection. For the entire teaching process, the deepest comprehension by students is that conceiving a pure form is quite easy, but it is not easy to work out a form that has mechanical characteristics of rational dimensions and full utilization of materials and that is easy for field construction, and that such form is indeed a real form.

If one course could survive for a long time, it implies that what studied in the course must be something with core value of architecture, and the Construction Technology Research is just such a course, which has been taught for 15 years since it was launched. Over the 15 years, the form and content of the course changed constantly along with change of external conditions, and such change is expected to happen again in the future, and we are also looking forward to such change, since construction has always been an activity happened along with changing conditions.

利用竹材容易弯曲的特性形成空间，结合已经成熟的编织技术，辅以简单的节点连接，适应山区人力、物力不足的条件状况，使得在较快的时间内建造游戏场所。

Make use of bamboo's bend property to form a space, combine with mature knitting techniques and supplement simple joint connection, to adapt to mountainous area in shortage of manpower and material resources, so that the game space can be built in a short time.

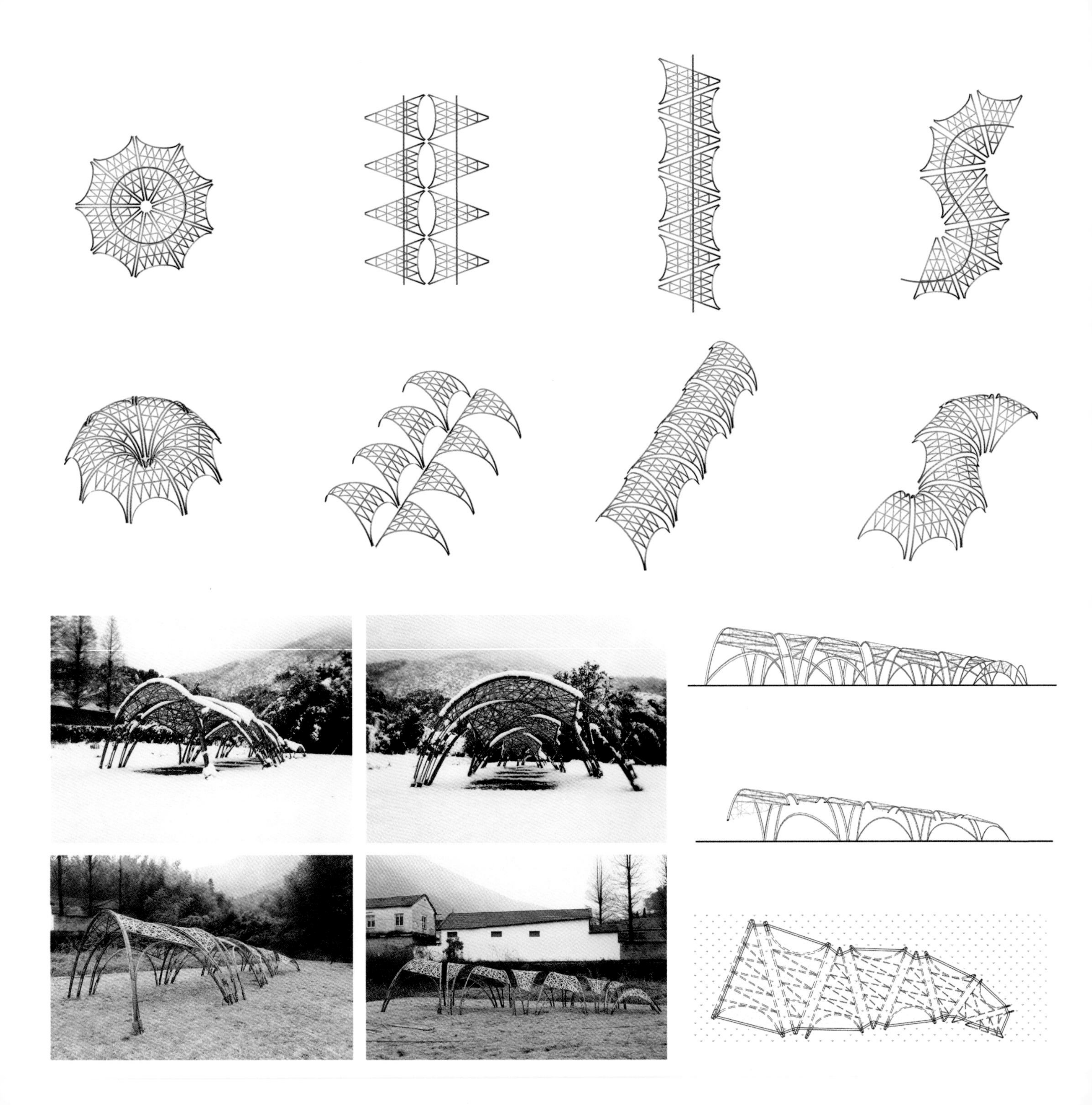

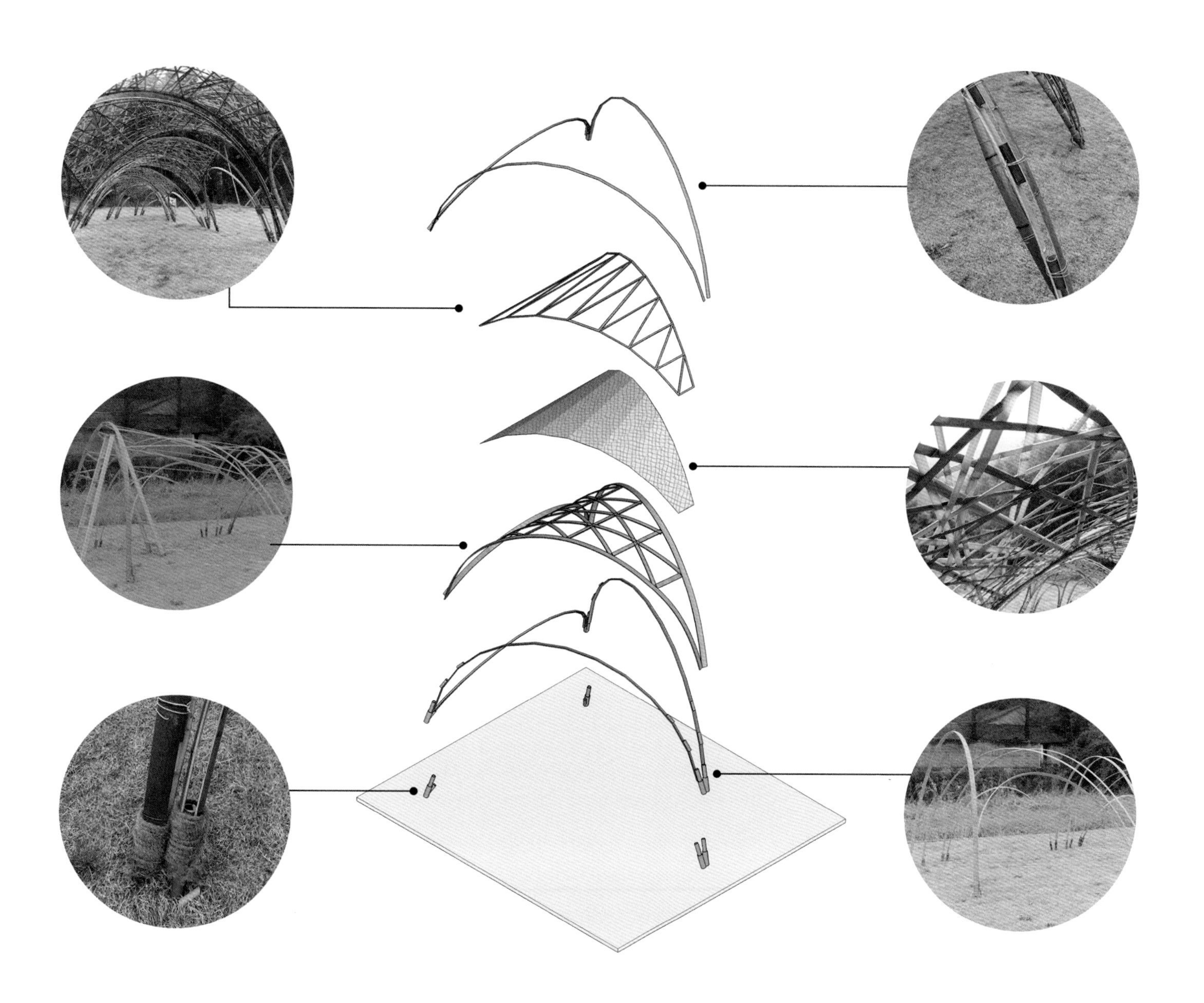

方案从竹子本身易弯的特性出发，通过弧形的基本构件来进行形体组合。

The scheme starts with the bend property of bamboo and carries out form composition with the arc basic components.

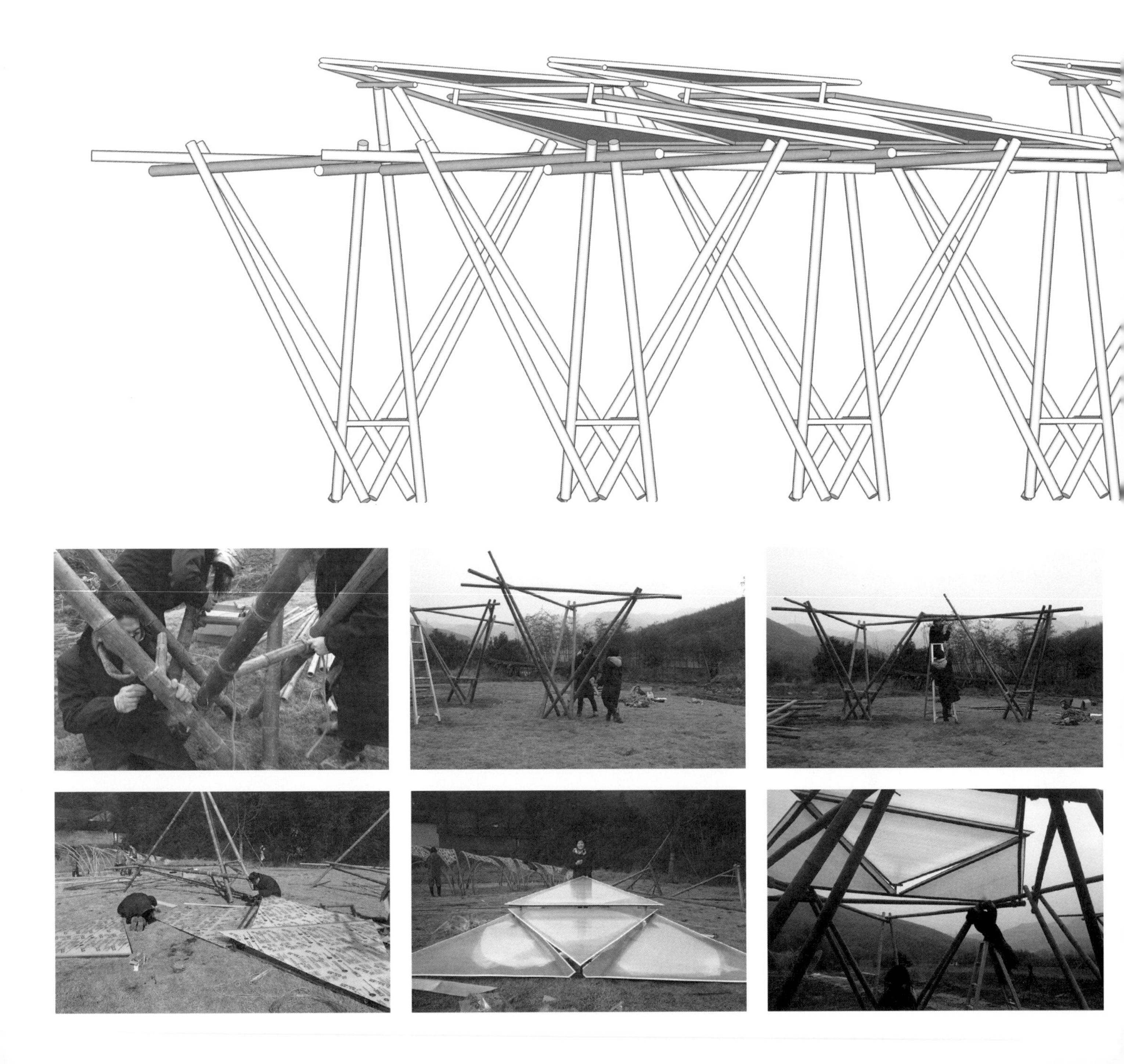

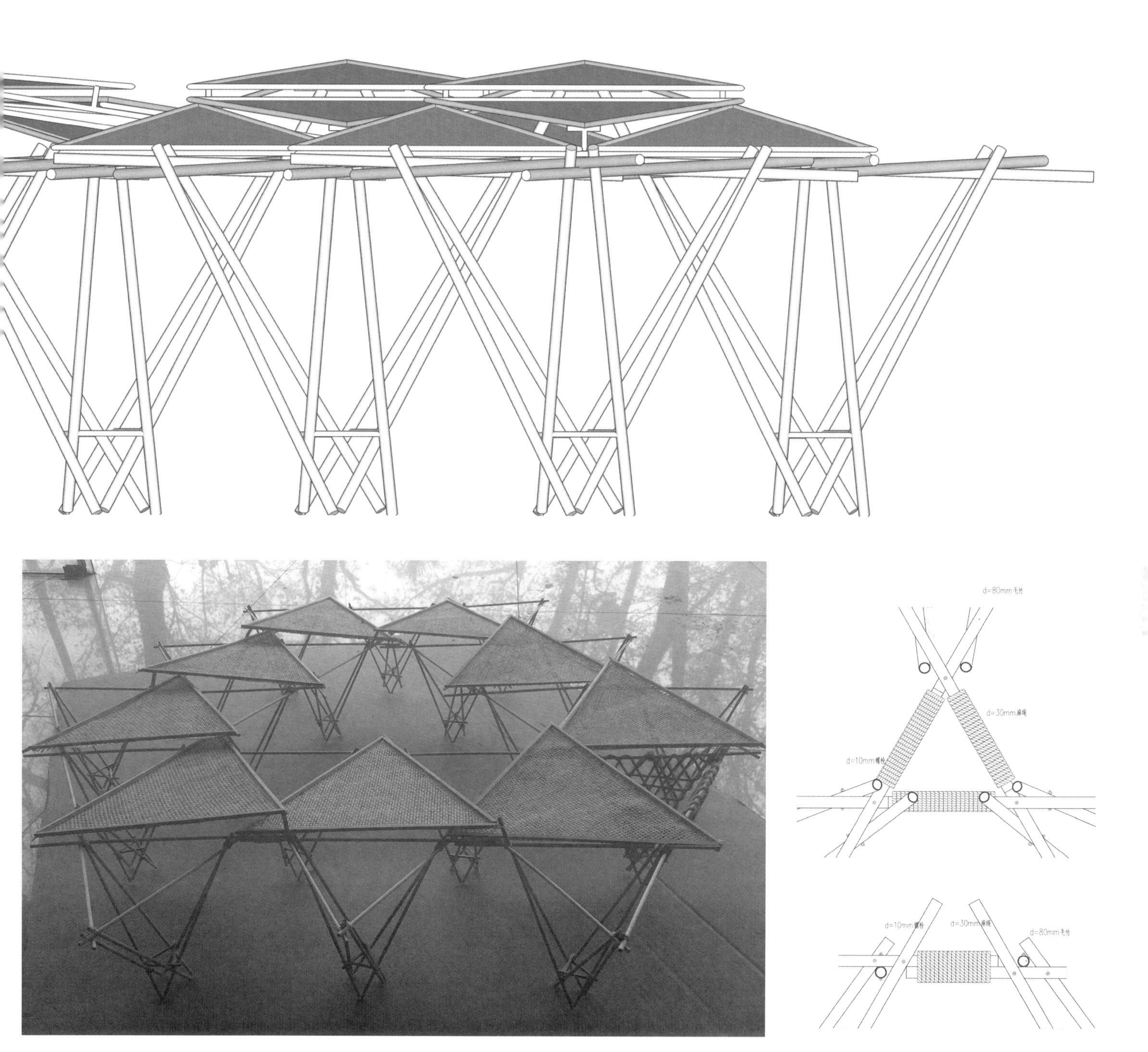
d=80mm 毛竹
d=30mm 麻绳
d=10mm 螺栓
d=10mm 螺栓
d=30mm 麻绳
d=80mm 毛竹

利用传统竹技术制作竹榫连接构建，设计了十二个节点连接。

Make bamboo tenon joints with traditional bamboo technology for connection and construction and design twelve joint connections.

建筑设计课程
ARCHITECTURAL DESIGN COURSES

本科一年级

设计基础（一）

· 季鹏

课程类型：必修

学时学分：64 学时／ 2 学分

Undergraduate Program 1st Year

BASIC DESIGN 1·JI Peng

Type: Required Course

Study Period and Credits: 64 hours/2 credits

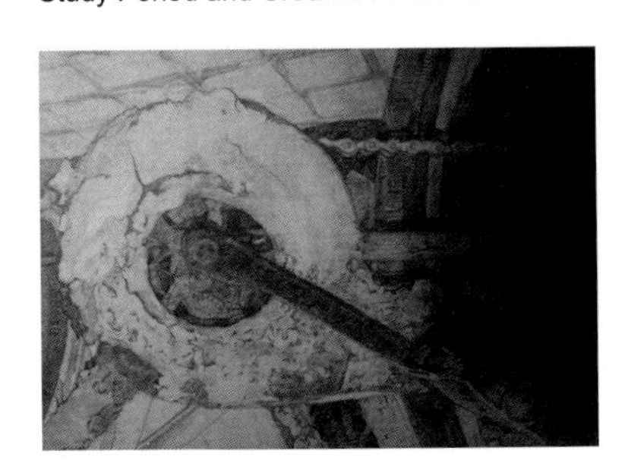

教学目标

提升学生感知美、捕捉美和创造美的能力。

研究主题

1.看到世界的美。

2.看到形式的美。

3.理解空间的规则。

4.理解使用的规则。

教学内容

1.研究对象的素描表达、黑白灰归纳、拼贴表现和陶土烧造。

2.针对Zoom in概念的训练与纸构成训练。

3.建筑摄影。

4.观念与材料的关系。

Training Objective

Improve ability of students to perceive, capture, and create beauty.

Research Subject

1. See beauty of the world.

2.See beauty of forms.

3.Understand rules of space.

4.Understand rules of use.

Teaching Content

1.Sketch presentation, black-white-grey induction, collage expression and pottery making for research objects.

2.Training on the concept of “Zoom in” and paper construction.

3.Architectural photography.

4.Relations between ideas and materials.

本科一年级

设计基础（二）

· 鲁安东　丁沃沃　唐莲

课程类型：必修

学时学分：64 学时／ 2 学分

Undergraduate Program 1st Year

BASIC DESIGN 2 · LU Andong, DING Wowo, TANG Lian

Type: Required Course

Study Period and Credits: 64 hours/2 credits

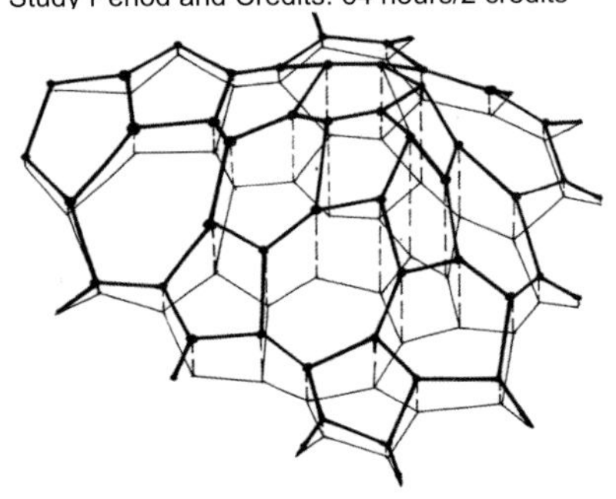

动作—空间分析

通过影像记录人在空间中的动作，挑选关键帧进行动作、尺度、几何与感知分析。目的在于：使学生初步认识身体、尺度与环境的相互影响；学会观察并理解场地；初步认识形式与背后规则的关系；学会发现日常使用中的问题并解决问题；学会使用分析图交流构思。

折纸空间

折叠纸板创造空间，用轴测图、拼贴图再现空间。目的在于：使学生初步掌握二维到三维的转化，初步认识图和空间的再现关系；利用单一材料围合复杂空间，初步认识结构与空间的关系；学会用分析图进行表述。

互承的艺术

运用互承结构原理，搭建人能进入的覆盖空间。目的在于：初步理解结构知识对于构筑空间的意义；在搭建过程中初步建立材料、节点、造价等概念；强化场地意识（包括朝向、环境、流线等）；学会用分析图进行表述。

Action-space Analysis

Record human’s actions in space with video and select key frames to conduct analysis on actions, dimensions, geometry and perception. It aims to enable students to preliminarily understand the mutual influence among body, dimensions and environment; learn how to observe and understand the site; preliminarily understand relations between forms and rules behind them; learn how to find out problems in daily use and how to solve these problems; learn how to use analysis charts to exchange ideas.

Folding Space

Fold paperboard to create space, and represent space with axonometric drawing and collage. It aims to enable students to preliminarily grasp the transformation from 2D to 3D, preliminarily understand the reappearance relationship between drawing and space; enclose complex space with single material, and preliminarily understand the relations between structure and space; learn how to express with analysis charts.

Art of Mutually-supporting

Apply principles of mutually-supporting structure to erect an accessible covered space. It aims to preliminarily understand meaning of structure knowledge to construction of space; roughly establish the concepts material, joints, and cost during erection; enhance site awareness (including orientation, environment, streamline, etc.); to learn expression with analysis charts.

本科二年级

建筑设计基础

· 刘铨　冷天

课程类型：必修

学时学分：64 学时／ 4 学分

Undergraduate Program 2nd Year

ARCHITECTURAL BASIC DESIGN · LIU Quan, LENG Tian

Type: Required Course

Study Period and Credits: 64 hours/4 credits

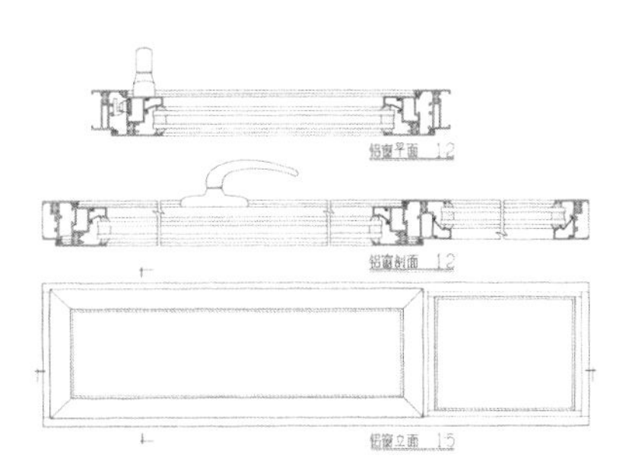

课题内容

认知与表达

教学目标

本课程是建筑学专业本科生的专业通识基础课程。本课程的任务主要是一方面让新生从专业的角度认知与实体建筑相关的基本知识，如主要建筑构件与材料、基本构造原理、空间尺度、建筑环境等知识；另一方面通过学习运用建筑学的专业表达方法来更好地掌握这些建筑基本知识，为今后深入的专业学习奠定基础。

教学内容

1.认知建筑

（1）立面局部测绘

（2）建筑平、剖面测绘

（3）建筑构件测绘

2.认知图示

（1）单体建筑图示认知

（2）建筑构件图示认知

3.认知环境

（1）街道空间认知

（2）建筑肌理类型认知

（3）地形与植被认知

4.专业建筑表达

（1）建筑图纸表达

（2）建筑模型表达

（3）环境分析图表达

Subject Content

Cognition and presentation

Training Objective

The course is the basic course of general professional knowledge for undergraduates of architecture. Task of the course is, on the one hand, allow students to cognize basic knowledge about physical building from an professional perspective, such as main building members and materials, basic constructional principles, spatial dimensions, and building environment; and on the other hand, to better master such basic architectural knowledge through studying application of professional presentation method of architecture, and to lay down solid foundation for future in-depth study of professional knowledge.

Teaching Content

1. Cognizing building

(1) Surveying and drawing of partial elevation

(2) Surveying and drawing plans, profiles of building

(3) Surveying and drawing building members

2. Cognizing drawings

(1) Cognition to drawings of individual building

(2) Cognition to drawings of building members

3.Cognizing environment

(1) Cognition to street space

(2) Cognition to types of building texture

(3) Cognition to terrain and vegetation

4. Professional architectural presentation

(1) Presentation with architectural drawings

(2) Presentation with architectural models

(3) Presentation with environmental analysis charts

本科二年级

建筑设计（一）：老城住宅设计

· 刘铨　冷天　王丹丹

课程类型：必修

学时学分：64 学时／ 4 学分

Undergraduate Program 2nd Year

ARCHITECTURAL DESIGN 1: RESIDENTIAL DESIGN OF THE DOWNTOWN · LIU Quan, LENG Tian,WANG Dandan

Type: Required Course

Study Period and Credits: 64 hours/4 credits

教学目标

本次练习的主要任务，是综合运用在建筑设计基础课程中的知识点来推进设计，初步体验用空间形式语言进行设计操作。在设定上，教案希望学生在设计学习开始之初，就关注场地条件、功能与空间、流线组织与人体尺度的紧密关系。

设计要点

1. 场地与界面：场地从外部限定了建筑空间的生成条件。本次设计场地是传统老城内真实的建筑地块，面积在 100 m^2 左右，单面或相邻两边临街，周边为 1~2 层的传统民居。主要要求学生从场地原有界面的角度来考虑设计建筑的形体、布局及其最终的空间视觉感受。

2. 功能与空间：本次设计的建筑功能设定为家庭住宅，建筑面积 100 m^2，建筑高度限制在 8 m（无地下空间）。

3. 流线组织：在给定场地内生成建筑，一方面，内部空间的安排要考虑到与场地界面的关系，如街道界面连续性、出入口的位置等。另一方面，空间的安排要考虑内部流线组织的合理性。

4. 人体尺度：在空间形式处理中注意通过图示表达理解空间构成要素与人的空间体验的关系，主要包括尺度感和围合感。

Training Objective

This exercise mainly aims to promote the design through comprehensive application of knowledge learned in the basic courses of architectural design, and to preliminarily experience design operation with spatial formal language. The teaching plan hopes that the students can pay attention to the close relationship among site conditions, function and space, streamline organization and the dimension of human figures at the beginning of design learning.

Key Points of Design

1. Site and interface: The site restricts the generating condition of architectural space from the outside. This design site is a real building plot in traditional downtown, covering an area of about 100 m^2, with frontage on one street or frontages on two streets, surrounded by traditional residences of 1~2 floors. The students are mainly required to consider and design the architectural form and structure, layout and final spatial visual perception from the perspective of original interface of the site.

2. Function and space: The architectural function of this design is defined as family house, with a floorage of 100 m^2 and a height of 8 m (without underground space).

3. Streamline organization: Generate a building in the given site and ensure that, on one hand, the arrangement of internal space takes into account its relationship with the site interface, such as the continuity of street interface, position of entrances and exits, etc.; on the other hand, the spatial arrangement takes into account the reasonability of the internal streamline organization.

4. Dimension of human figures: In the processing of space form, understand the relationship between the space elements and spatial experience of human through graphic expression, mainly including the sense of dimensions and the sense of closure.

本科二年级

建筑设计（二）：小型公共建筑设计

·刘铨　冷天　王丹丹

课程类型：必修

学时学分：64 学时／4 学分

Undergraduate Program 2nd Year

ARCHITECTURAL DESIGN 2: SMALL PUBLIC BUILDING · LIU Quan, LENG Tian,WANG Dandan

Type: Required Course

Study Period and Credits: 64 hours/4 credits

课题内容

风景区茶室设计

教学目标

本课程承接上一个设计题目，继续训练使用空间形式语言进行设计操作，训练学生进一步掌握设计的“空间形式语言”。

设计要点

场地与界面，功能与空间，空间与组织，尺度与感知。

教学内容

1. 场地与界面：场地从外部限定建筑空间的生成条件，要求学生从场地水平向界面的限定来考虑设计建筑的形体、布局及其最终的空间视觉感受。

要求坡地界面形态加以形式化的提取和表达，如堆叠、阵列、编制、切片、张拉等，以强化对地形的理解并作为推动后续设计的形式工具。

2. 功能与空间：建筑面积300 m^2，建筑层数2 层（无地下空间），其中包括入口门厅、容纳60 人的大厅、容纳30 人的2~4 人雅座若干，以及操作间、储藏间、值班室、卫生间等必要的辅助空间。

3. 空间的组织：基于地形的形式化表达，寻求适应建筑空间需求的场地改造形式进行高差的处理，把对地形的理解与公共建筑的基本功能空间组织模式及形态塑造联系起来。

4. 尺度与感知：在空间形式处理中注意通过图示表达理解空间构成要素与人的空间体验的关系，主要包括尺度感和围合感，注意景观朝向问题。

Subject Content

Design of Teahouse in Scenic Spot

Training Objective

This course continues from last design subject, proceeds with the training of design operation with spatial formal language and trains the students to further master the “spatial formal language” for design.

Key Points of Design

Site and interface, function and space, organization of space, dimensions and perception.

Teaching Content

1. Site and interface: The site restricts the generating condition of architectural space from the outside. The students are required to consider and design the architectural form and structure, layout and final spatial visual perception from the perspective of horizontal interface of the site.

In addition, the students are required to make formalized extractions and expressions of slope interface morphology, such as overlapping, arraying, forming, sectioning, tensioning, etc., to strengthen the understanding of terrain and use them as form tools to promote follow-up design.

2. Function and space: The building covers 300 m^2, has two floors (without underground space) and includes an entrance hallway, a hall that can accommodate 60 people, a number of 2~4-seat booths that can accommodate 30 people, as well as the operation room, storeroom, duty room, rest room and other necessary auxiliary space.

3. Space organization: Based on the formalized expression of terrain, seek for a site transformation form meeting the architectural space requirement to process the height difference, and combine the understanding of the terrain with spatial organization pattern and form modeling of basic functions of public buildings.

4. Dimensions and perception:In the processing of space form, understand the relationship between the space elements and spatial experience of human through graphic expression, mainly including the sense of dimensions and the sense of closure. Pay attention to the orientation of landscape.

本科三年级

建筑设计研究（三）：小型公共建筑设计

·周凌　童滋雨　窦平平

课程类型：必修

学时学分：72 学时／4 学分

Undergraduate Program 3rd Year

ARCHITECTURAL DESIGN 3: SMALL PUBLIC BUILDING · ZHOU Ling, TONG Ziyu,DOU Pingping

Type: Required Course

Study Period and Credits: 72 hours/4 credits

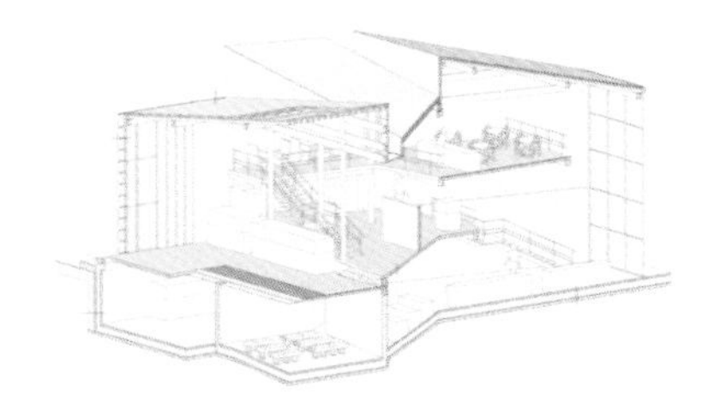

课题内容

赛珍珠纪念馆扩建

教学目标

此课程训练最基本的建造问题，使学生在学习设计的初始阶段就知道房子如何造起来，深入认识形成建筑的基本条件：结构、材料、构造原理及其应用方法，同时课程也面对场地、环境和功能问题。训练核心是结构、材料、场地。在学习组织功能与场地同时，强化认识建筑结构、建筑构件、建筑围护等实体要素。

文脉：充分考虑校园环境、历史建筑、校园围墙以及现有绿化，需与环境取得良好关系。

退让：建筑基底与投影不可超出红线范围。若与主体或相邻建筑连接，需满足防火规范。

边界：建筑与环境之间的界面协调，各户之间界面协调。基底分隔物（围墙或绿化等）不超出用地红线。

户外空间：扩建部分保持一定的户外空间，户外空间可在地下。

地下空间：充分利用地下空间。

教学内容

基地内地面最大可建面积约 100 m^2，地下可建面积 200~300 m^2，总建筑面积约 400~500 m^2，建筑地上 1 层，限高 6 m，地下层数层高不限，展示区域 200~300 m^2，导游处 10 m^2，纪念品部 30 m^2，茶餐厅 60 m^2，厨房区域 >10 m^2，另包括 门厅与交通、卫生间。

Subject Content

Extension of Pearl S.Buck's House in Nanjing

Training Objective

This course trains the students to solve the basic construction of architecture. Students should learn how to build an architecture at the very beginning of their studying, understand the basic aspects of architectures: the principles and applications of structure, material and construction. The course also includes the problem of site, enviroment and function. The keypoints of the course include site, structure and material. Students should strengthen the understanding of physical elements including structures, components and facades while learning to organize the function and site.

Context: The enviroment, historical building, the edge of the campus and the green belt around the site should be taken into consideration. The expansion is expected to have a good relationship with the surroudings.

Retreat Distance: The new architecture can't beyond the red line. Fire protection rule should be complied.

Boundary: Both the boundary between different buildings and between building and environment should be harmonized.

Open Space: Open space should be considered, which permitted to be placed underground.

Underground Space: Underground space should be well used.

Teaching Content

The maximum ground can be used in the base area is about 100 m^2, while underground construction area is about 200~300 m^2, and the total floor area of architecture should be about 400~500 m^2.The architecture should be 1 floor above the groud lower than 6 m.The underground levels have no limitation.Exhibition area: 200~300 m^2, information center: 10 m^2, shop: 30 m^2, tea cafe: 60 m^2, kitchen: >10 m^2, lobby & walking space, toilet.

本科三年级

建筑设计（四）：中型公共建筑设计

· 周凌　童滋雨　窦平平

课程类型：必修

学时学分：72 学时 / 4 学分

Undergraduate Program 3rd Year

ARCHITECTURAL DESIGN 4: PUBLIC BUILDING · ZHOU Ling, TONG Ziyu,DOU Pingping

Type: Required Course

Study Period and Credits: 72 hours/4 credits

课题内容

傅抱石美术馆

教学目标

课程主题是“空间”和“流线”，学习建筑空间组织的技巧和方法，训练空间的效果与表达。空间问题是建筑学的基本问题，课题基于复杂空间组织的训练和学习。从空间秩序入手，安排大空间与小空间、独立空间与重复空间，区分公共与私密空间、服务与被服务空间、开放与封闭空间。同时，空间的串联形成序列，需要有效组织流线，并且充分考虑人在空间中的行为和空间感受。以模型为手段，辅助推敲。设计阶段分体积、空间、结构、围合等，最终形成一个完整的设计。

教学内容

1.空间组织原则：空间组织要有明确特征，有明确意图，概念要清楚，并且满足功能合理、环境协调、流线便捷的要求。注意三种空间：聚散空间（门厅、出入口、走廊）；序列空间（单元空间）；贯通空间（平面和剖面上均需要贯通，内外贯通、左右前后贯通、上下贯通）。

2.空间类型：展览陈列空间：3000 m^2；收藏保管空间：700 m^2；技术、研究空间：240 m^2；行政办公空间：150 m^2；休闲服务空间：300 m^2；其他空间：传达室10 m^2，设备房200 m^2，交通门厅面积自定，客用、货用电梯各1部，室外停车场。

建筑面积不超过5000 m^2，高度不超过18 m。

Subject Content

Fu Baoshi Art Gallery

Training Objective

The course topic is space and circulation, learning techniques and methods of architectural space organization, and training on effect and presentation of space. Space is a basic issue for architecture, and the course is based on training and study on organization of complex spaces. Start from spatial order to arrange large and small spaces, independent space and overlapped space, and to distinguish public and private spaces, serving and served spaces, open and closed spaces. Meanwhile, linking spaces to shape sequence requires effective organizational circulation, as well as full consideration of behaviors, spatial feeling of people in space. Use models as means to assist deliberation. Design stages include volume, space, structure, and enclosure, and shape a complete design in the end.

Teaching Content

1.Space organization principles: Space organization requires distinctive characteristics, explicit intention, and clear concepts. It shall also meet the requirements of reasonable functions, coordinated environment, and convenient circulation. Attention shall be paid to three types of space: converging and diverging space (hallway, entrance and exit, corridor); sequence space (unit space); connecting space (connecting spaces are required on plans and profiles, internal-and-external connection, left-and-right, front-and-rear connections, up-and-down connection).

2.Space type: exhibition & showcase space: 3000 m^2; collection & storage space: 700 m^2; technical, research space: 240 m^2; administrative office space: 150 m^2; leisure service space: 300 m^2; other spaces: janitor's room 10 m^2, equipment room 200 m^2, area of traffic hallway to be determined, one guest elevator and one goods elevator, outdoor parking lot.

The floor area shall not exceed 5000 m^2, and the height shall not exceed 18 m.

本科三年级

建筑设计（五—六）：大型公共建筑设计

· 华晓宁　钟华颖　王铠

课程类型：必修

学时学分：144 学时 / 8 学分

Undergraduate Program 3rd Year

ARCHITECTURAL DESIGN 5-6: COMPLEX BUILDING · HUA Xiaoning, ZHONG Huaying, WANG Kai

Type: Required Course

Study Period and Credits: 144 hours/8 credits

课题内容

城市建筑——社区商业中心＋活动中心设计

研究主题

1. 实与空：关注城市中建筑实体与空间的相互定义、相互显现，将以往习惯上对于建筑本体的过度关注拓展到对于“之间”的空间的关注。

2. 内与外：进一步突破“自身”与“他者”之间的界限，将个体建筑的空间与城市空间视为一个连续统，建筑空间即城市空间的延续，城市空间亦即建筑空间的拓展，两者时刻在对话、互动和融合。

3. 层与流：不同类型的人和物的行为与流动是所有城市与建筑空间的基本框架，当代大都市中不同的流线在不同的高度上层叠交织，构成一个复杂的多维城市。必须首先关注行为和流线的组织，由此才生发出空间的系统和形态。

4. 轴与界：城市纷繁复杂的形态表象之后隐含着秩序和控制性，并将成为新的形态介入。

教学内容

在用地上布置社区商业中心（约 15000 m^2）、社区文体活动中心（约8000 m^2），并生成相应的城市外部公共空间。

Subject Content

Urban Buildings—Design of Community Business Center and Activity Center

Research Subject

1. Entity and space: Pay attention to mutual definition, mutual representation of architectural entity and space in cities, and extend traditional excessive attention to the building itself to the space “among them”.

2. Interior and exterior: Further break through the boundary between “self” and “others”, and consider space of individual buildings and urban space as a continuum.

3. Stack and flow: Behaviors and flows of different types of people and objects are the basic framework of all urban and building spaces, and a complex multi-dimensional city is formed by stacking up and interweaving of different flow lines at different altitudes in modern metropolis. We must pay attention to organization of behaviors and flow lines first, and then can generate system and morphology of space.

4. Axis and boundaries: Order and control are concealed behind the morphologic appearance of complexity of cities, which will be involved as new forms.

Teaching Content

Lay out community commercial center (about 15000 m^2) and community recreational and sports activities center (about 8000 m^2) on the land, and generate associated outdoor urban public spaces.

本科四年级

建筑设计（七）：高层建筑设计

· 吉国华　胡友培　尹航

课程类型：必修

学时学分：72 学时／ 4 学分

Undergraduate Program 4th Year

ARCHITECTURAL DESIGN 7: HIGH-RISING BUILDING · JI Guohua, HU Youpei, YIN Hang

Type: Required Course

Study Period and Credits: 72 hours/4 credits

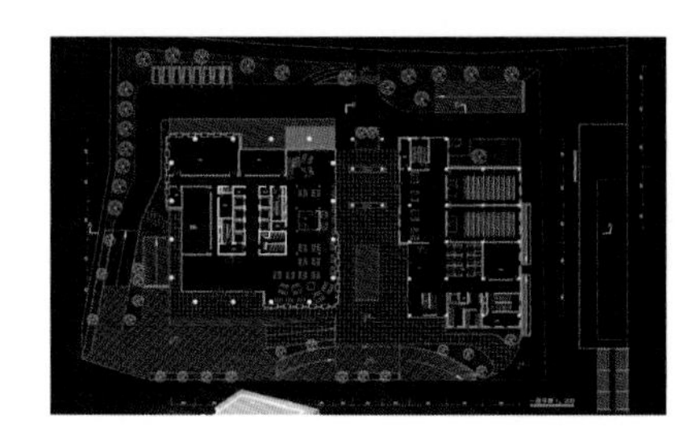

课题内容

高层办公楼设计

教学目标

高层办公建筑设计涉及城市、空间、形体、结构、设备、材料、消防等方面内容，是一项较复杂与综合的任务。本课题采取贴近真实实践的视角，教学重点与目标是帮助学生理解、消化以上各方面知识，提高综合运用并创造性解决问题的技能。

教学内容

建筑容积率≤5.6，建筑限高≤100 m，裙房高度≤24 m，建筑密度≤ 40%。需规划合理流线，避免形成交通拥堵。

高层部分为办公楼，设计应兼顾各种办公空间形式。裙房设置会议中心，须设置 400 人报告厅 1 个，200 人报告厅 2 个，100 人报告厅 4 个，其他各种会议形式的中小型会议室若干，以及咖啡茶室、休息厅、服务用房等。会议中心应可独立对外使用。机动车交通独立设置，不得进入校内道路系统。地下部分主要为车库和设备用房。

Subject Content

Design of High-rise Office Building

Training Objective

Design of the high-rise office building is a complicated and comprehensive task, involving city, space, form, structure, equipment, materials and fire control etc. From a perspective close to the practice, this course focuses on and aims at helping students understand and grasp the knowledge of the above-mentioned aspects and improving their skills of integrated use and creatively solving problems.

Teaching Content

Building plot ratio ≤ 5.6, building height limit ≤ 100 m, annex height ≤ 24 m, building density ≤ 40%. Reasonable circulation must be planned to avoid traffic jam.

The high-rise part is an office building, so the design must give consideration to various forms of office space.The annex is a conference center, which must include 1 lecture hall of 400 seating capacity, 2 lecture halls of 200 seating capacity, 4 lecture halls of 100 seating capacity, several small and medium-sized meeting rooms for various meetings, and coffee & tea room, lobby, and service quarter etc. The conference center shall be separated and available for external usage. Motor vehicle traffic routes must be separated from road system within the campus. The underground part is mainly for garage and equipment room.

本科四年级

建筑设计（八）：城市设计

· 吉国华　胡友培　尹航

课程类型：必修

学时学分：72 学时／ 4 学分

Undergraduate Program 4th Year

ARCHITECTURAL DESIGN 8: URBAN DESIGN · JI Guohua,HU Youpei, YIN Hang

Type: Required Course

Study Period and Credits: 72 hours/4 credits

课题内容

南京碑亭巷地块旧城更新城市设计

教学目标

1.着重训练城市空间场所的创造能力，通过体验认知城市公共开放空间与城市日常生活场所的关联，运用景观环境的策略创造城市空间的特征。

2.熟练掌握城市设计的方法，熟悉从宏观整体层面处理不同尺度空间的能力，并有效地进行图纸表达。

3.理解城市更新的概念和价值；通过分析理解城市交通、城市设施在城市体系中的作用。

4.多人小组合伙，培养团队合作意识和分工协作的工作方式。

教学内容

1.设计地块位于南京市玄武区，总用地约为6.20 hm^2。地块内国民大会堂旧址、国立美术陈列馆旧址和北侧邮政所大楼可保留，其余地块均需进行更新。地块周边有丰富的博物馆、民国建筑等文化资源，设计应对周边文化环境起到进一步提升作用。地块周边用地情况复杂，设计中需考虑与周边现状的相互影响。

2.本次设计的总容积率指标为2.0~2.5，建筑退让、日照等均按相关法规执行。

3.碑亭巷、石婆婆庵需保留，碑亭巷和太平北路之间现状道路可根据设计进行位置或线形调整。

4.地下空间除满足单一地块建筑配建的停车需求外，应综合考虑地上、地下城市一体化设计与综合开发。

Subject Content

Urban Design for Old Town Renovation of the Land Parcel of Beiting Lane in Nanjing

Training Objective

1. Emphasize the training on ability of creating urban spatial places, and create features of urban space with the strategy of landscape environment through experiencing and perceiving the links between urban public spaces and urban daily living places.

2. Master methodology of urban design, grasp the ability of handling spaces of different dimension at macro and integral level, and achieve effective representation with drawings.

3. Understand the concept and value of urban renovation; understand the role of urban traffic, urban facilities in the urban system through analysis.

4. Form partnership with several group members to cultivate awareness of teamwork and the working mode of collaboration.

Teaching Content

1.Land parcel of the design is located in Xuanwu District, Nanjing, covering an area of 6.20 hm^2 approximately. Site of the former National Assembly Hall, site of the former National Art Gallery and the post office building at north side may be retained, other parts of the land parcel need to be renovated. There are abundant cultural resources such as museums and buildings constructed in the period of the Republic of China around the land parcel, so the design should further improve the cultural environment around it. Land use conditions around the land parcel is very complicated, so mutual influence with surrounding existing conditions must be taken into account in the design.

2.Gross plot ratio of the design is 2.0~2.5, and building setback and sunlight value shall comply with relevant laws and regulations.

3.Beiting Lane and Shipopo Nunnery are to be retained, and location and route of the existing road between Beiting Lane and North Taiping Road may be adjusted according to the design.

4.For underground space, besides meeting the associated parking demand of buildings on the single land parcel, overall consideration shall be taken for integrated design and comprehensive development of urban spaces above and under the ground.

本科四年级

毕业设计

· 赵辰

课程类型：必修

学时学分：1 学期 /0.75 学分

Undergraduate Program 4th Year

THESIS PROJECT · ZHAO Chen

Type: Required Course

Study Period and Credits: 1 term /0.75 credit

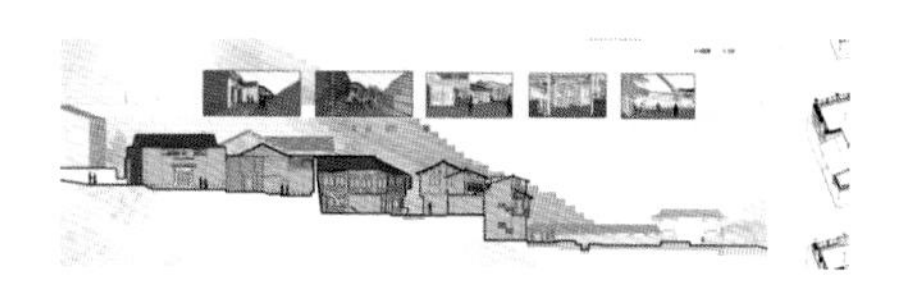

课题内容

闽东北山区低碳生态型度假村落项目规划设计

课题介绍

锦屏位于福建东北部山区的南平市政和县，是生态、人文资源都十分优秀的地域，因交通缘故而长期受到发展的限制。随着高速公路与铁路的发展，这一地区将迅速得到新的发展机遇，而新的发展必须适应新的低碳、生态型的发展模式，同时充分保护与发挥地域历史文化与生态的优势。这正是新时代对建筑与乡村规划设计的一种挑战。

教学目标

掌握建筑设计基本的技能与知识（测绘、建模、调研、分析），并能对特定的地域和历史建筑进行深入的设计研究（内容策划、建筑结构和构造），根据社会发展的需求，提出改造和创造的可能。在选定的村落之现状研究的基础上，进行村落景观空间的整体规划，并选择相关重点区域与建筑，进行专项的建筑设计。

Subject Content

Design and Arrangement of Low Carbon and Ecological Village in Northeastern Mountain Area of Fujian Province

Subject Description

Jinping lies in Zhenghe County, Nanping, northeastern mountain area of Fujian province. It is a place where both ecological resources and human resources are abundant. However, because of traffic, the development of this area is restrained. Accompanying development of expressways and railways, this area will rapidly have a new chance to develop, but the new development must adapt to new, low carbon and ecological way to develop. Meanwhile, advantages of historical culture and ecology should be well protected and promoted. This is exactly a challenge, which is produced by new time, of architecture and rural planning and designing.

Teaching Objective

Master basic skills and knowledge (surveying and mapping, modeling, investigation, analysis) of architecture design; have profound design and study (content planning, structure of architecture, construction) of special area and historical buildings; put forward possibility of designing and creating, according to needs and requirements of social development. Based on the study of present situation of appointed villages, plan the whole village landscape. Select related key area and buildings, and design special buildings.

本科四年级

毕业设计

· 丁沃沃　胡友培

课程类型：必修

学时学分：1 学期 /0.75 学分

Undergraduate Program 4th Year

THESIS PROJECT · DING Wowo,HU Youpei

Type: Required Course

Study Period and Credits: 1 term /0.75 credit

课题内容

福建长汀历史文化名城：城市更新与建筑设计

课题介绍

本课题以长汀古城历史街区为设计范围，通过调研和访谈理解设计问题。通过测绘和分析学习传统建筑的类型和优势，以及地方建造的工法。阅读文献资料相关理论，并通过具体的设计研究与实验将所学转化为学理层面的知识和设计方法。

教学目标

基于历史文化名城保护与更新的城市设计的真实项目，本毕业设计了涵盖历史文化知识、典型民居类型、建筑设计与建造和城市设计方法的等训练计划，旨在通过训练学习科学有效的调研与分析方法、多重限定下的建筑设计、面向建造的真实问题和城市设计的现实意义。在将本科所学知识融会贯通的基础上，理解设计与研究的关系和研究对于设计的价值。

Subject Content

Famous Historical and Cultural City of Changting, Fujian: City Renovation and Architecture Design

Subject Description

This project takes the historical street block of this ancient city-Changting as its design sphere, through the investigation and interviews to learn about the issue of design and through the surveying and mapping to learn the traditional architecture types and advantages as well as techniques of local construction. By reading relevant literature materials, and through the specific design research and experiments, this project aims to transfer the knowledge learnt in the past into academic knowledge and design methodology.

Teaching Objective

Based on the actual program of urban planning of the protection and renovation of the famous historical and cultural city, this graduation project covers a series of training plans including historical and cultural knowledge, typical folk house types, architecture design and construction, and urban designing methods, etc. in a bid to learn about scientific and effective investigation and analysis methods, architecture design under multiple restrictions, actual problems in construction and practical significance of urban planning. On the basis of integrating and combing the knowledge of the undergraduate courses, this project aims to gain understanding of the relationship between design and study and the value of the study for the design.

本科四年级

毕业设计

· 冯金龙　郜志

课程类型：必修

学时学分：1 学期 /0.75 学分

Undergraduate Program 4th Year

THESIS PROJECT · FENG Jinlong,GAO Zhi

Type: Required Coure

Study Period and Credits: 1 term /0.75 credit

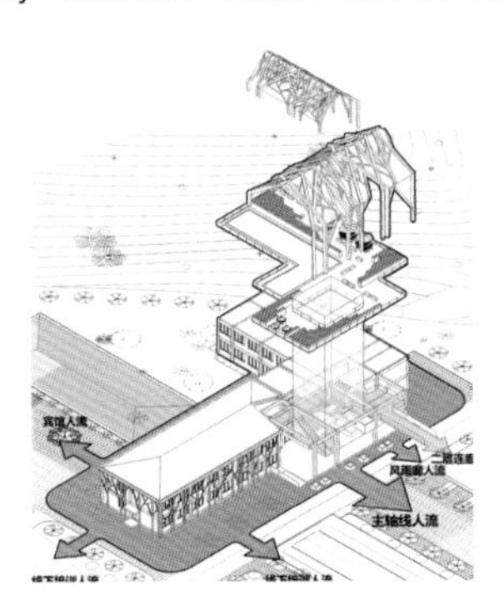

课题内容

建筑技术科学专门化设计 / 研究：雨花软件培训基地规划设计

课题介绍

基于可持续发展的目标，绿色建筑设计是未来建筑学研究中的重要方向。在建筑设计过程中绿色、低碳、节能与可持续的理念应作为重要的设计因素，与功能与形式更好地融为一体。将绿色建筑技术、城市微气候设计及建筑环境质量控制与建筑形态和空间生成进行深入整合是一个尤为重要的主题。雨花软件培训基地建设地点在南京雨花台区铁心桥街道马家店组，是雨花台区区委、区政府重点打造的创新创业载体和人才集聚高地，是雨花台区高科技产业发展、科技自主创新的先导区。其规划指导思想应贯穿“低碳、环保、可持续发展”的设计理念，为此，需要在建筑形态和空间的设计中深入整合相关绿色建筑技术，以使建筑既能满足使用要求，同时又能有效降低能耗，改善建筑环境质量。

Subject Content

The Specialized Design and Study of the Building Technology Science: The Planning and Design of Yuhua Software Training Base

Subject Description

Based on the goal of sustainable development, green architecture design is an important direction of the future research of architecture. In the process of architecture designing, the philosophy of green, low-carbon, energy conservation and sustainability should be the important designing factors and be integrated with the function and form in a batter fashion. Conducting an in-depth integration and combination of green architecture technology, urban micro climate design as well as quality control of architecture environment and architecture form and space is an utmost theme. Yuhua Software Training Base, located in Majiadianzu, Tiexin Briage Rd., Yuhuatai District, Nanjing, is a platform of innovation and entrepreneurship and a highland gathering various talents, built by the CPC District committee and the District Government. Meanwhile, it is also a leading area for hi-tech industry development and science and technology independent innovation of the Yuhua District. Its guiding principle should be based on the philosophy of low-carbon, environment-friendly and sustainable development. Therefore, the related green architecture technology should be deeply incorporated into the designing of the architecture form and space, so as to reduce energy consumption and improve the quality of architecture environment while meeting the requirement of utilization.

本科四年级

毕业设计

· 钟华颖

课程类型：必修

学时学分：1 学期 /0.75 学分

Undergraduate Program 4th Year

THESIS PROJECT · ZHONG Huaying

Type: Required Coure

Study Period and Credits: 1 term /0.75 credit

课题内容

基于张拉整体结构的数字化搭建

课题介绍

富勒发明的张拉整体结构，是一组不连续的受压构件与一套连续的受拉单元组成的自支承、自应力的空间网格结构。这一结构系统最大限度地利用了材料和截面的特性，用最少的材料营造了最大的跨度和空间。传统的张拉整体结构受压的刚性杆位于系统中央，占据了使用空间，长期以来较少用于建筑设计，多用于桥梁、雕塑。改进的张拉整体结构将结构杆与索置于周边，围合出使用空间，具有了应用于建筑设计的可能性。

张拉整体结构提供了一种结构原型，三维打印技术则可以灵活地制造建筑构件，为建筑围护结构的设计建造提供了新的手段。张拉整体结构与三维打印结合成为一种建筑加工建造的系统。

本设计拟在校园内选择某处闲置的场地，搭建一个具有一定使用空间的构筑物，如休息亭、坐椅、雨篷等。利用数字化手段设计结构形态及外围护系统，最终完成全尺寸模型的搭建，从而探索这一系统建筑应用的可能性。

Subject Content

Digital Construction Based on Tensegrity Structure

Subject Description

Tensegrity structure was invented by R.B.Fuller. It is a self-supporting and self-stressing space grid structure which consists of a series of uncontinuous compressed members and a series of continuous tension units. This structure can make fullest use of features of materials and cross section, and it has the biggest span and space with the least materials. Compressed rigid robe of traditional tensegrity structure locates in the center of the system and occupies some using space. So, for a long time, it is usually not used to architecture, but used in bridge and sculpture very often. Rigid robe and rope are put around the whole structure in modified tensegrity structure. Using space is surrounded by rigid robe and rope, so it becomes possible to be used into architecture.

Tensegrity structure offers a structure prototype. 3D printing technology can produce building units flexibly and becomes a new way to design and construct building envelope. Combination of tensegrity structure and 3D printing technology becomes a system of processing and constructing buildings. This design plans to choose an idle place in campus and construct a building that has some using space, such as lounge hall, bench, rain-shed and so on. Configuration and peripheral protection system are designed by digital means, and the construction of full-scale model is finished finally to explore possibility of application of this construction system.

本科四年级

毕业设计：基于机器人技术的墙体搭建

· 童滋雨

课程类型：必修

学时学分：1 学期 /0.75 学分

Undergraduate Program 4th Year

THESIS PROJECT: WALL CONSTRUCTION BASED ON ROBOTICS · TONG Ziyu

Type: Required Coure

Study Period and Credits: 1 term /0.75 credit

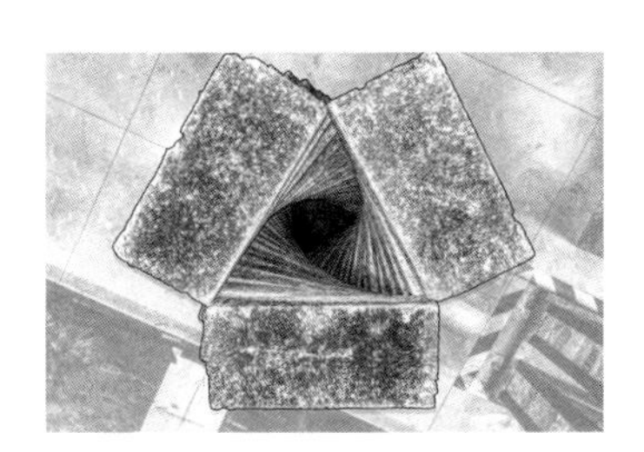

课题内容

机器人技术在数字化设计和建造中的应用

教学目标

建造一段 5 m 长、2.5 m 高的墙体，以标准大小的木砖不用胶水直接搭接而成。

要求和限定

1. 墙体的设计需在 Rhino 和 Grasshopper 平台上进行，具有参数化设计特征。

2. 墙体的搭建需使用机器人设备。

成果

1. 图纸：内容包括设计墙体的平立面图，生成程序的流程图，机器人设备的应用分析。

2. 模型：1:1 实物模型。

3. 研究文本：A4 幅面研究文本，内容包括设计逻辑解析、生成程序解读、机器人设备应用分析三个部分。设计逻辑解析是利用图示与文字说明对设计当中采取的各种规则进行分析说明，表明规则的合理性。生成程序解读是利用程序流程图，解释实现设计逻辑的计算机程序。机器人设备应用分析是对机器人在墙体搭建过程中应用方法和实现策略的分析。

Subject Content

Application of Robotics in digitization design and construction

Training Objective

Construct a 5 m L x 2.5 m H wall through lapping with standard sized wood bricks without glue.

Requirement and limit

1. The design of wall shall be done on the Platforms Rhino and Grasshopper, which shall be with parametrization design characteristics.

2. Construction of wall needs robot equipment.

Achievement

1. Drawing: the content includes the plan and elevation of the designed wall and the flow chart used for generating the programs and the applicable analysis of robot equipment.

2. Model: 1:1 physical model

3. Research text: A4 size research text's content includes design logic analysis, generating program understanding and robot equipment application analysis. Design logic analysis means analyzing and indicating all rules taken during the design by using of graphical representation and explanatory notes, and it means the rationality of rules. The generating program understanding interprets the computer program realizing design logic by using of program flow chart. Robot equipment application analysis means analyzing the application methods and realization strategy during the process of wall construction by robot.

本科四年级

毕业设计

· 周凌

课程类型：必修

学时学分：1 学期 /0.75 学分

Undergraduate Program 4th Year

THESIS PROJECT · ZHOU Ling

Type: Required Coure

Study Period and Credits: 1 term /0.75 credit

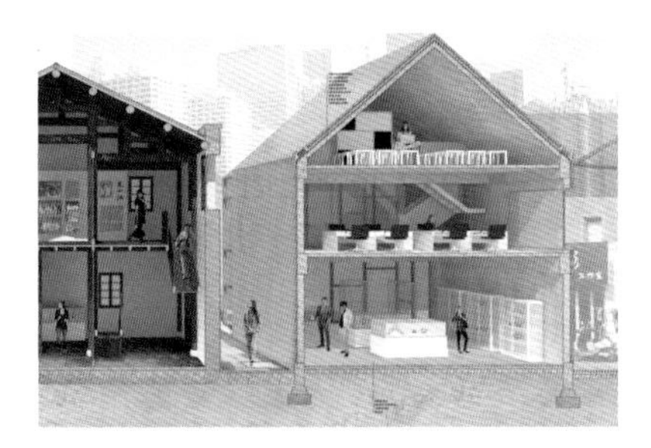

课题内容

乡村再造

教学目标

中国传统村镇正在消失，其速度与中国城市化速度成正比，传统农耕文化、传统手工艺、传统价值观处于消失弱化的边缘。本课题着重研究中国传统村镇保护更新的议题。学生通过调研和规划设计，了解传统村镇与传统建筑文化，学习规划知识，训练建筑设计技巧。

课题介绍

毕业设计将以武进杨桥古村或南京江宁苏家村的乡村营建为主题进行调研、测绘、改造设计与实践（选择其中一个题目实施）。

题目一：常州武进杨桥古村保护更新修建性规划设计；

题目二：南京江宁秣陵街道苏村文创市集与精品酒店改造设计（真题）。

重点问题

1. 村落自然与人文环境
2. 民居类型
3. 保护规划
4. 改造策划
5. 技术与建造

Subject Content

Reconstruction of Villages

Training Objective

Traditional Chinese villages are disappearing, and the disappearing speed is proportionate to the speed of urbanization. Traditional farming culture, traditional handicraft, traditional values are also endangered. This project puts more emphasis on studying the proposal of protecting and renewing Chinese traditional villages. Students gradually comprehend traditional villages and traditional architectural culture, learn knowledge of planning and gain skills of architectural design by research, design and planning.

Subject Description

This graduation project will focus on ancient village in Yangqiao, Wujin or rural management and construction of Su Village, Jiangning, Nanjing to do research, surveying and drawing, reconstruction design and practice (Only one of the themes should be chosen to do).

Theme one: planning and design of protecting, renewing and constructing ancient village in Yangqiao, Wujin, Changzhou.

Theme two: reconstruction design of cultural and creative bazaar and excellent hotel in Su Village, Moling Street, Jiangning, Nanjing.

Key Questions

1. Natural and humanistic environment of village
2. Types of folk house
3. Protection planning
4. Reconstruction plan
5. Technology and construction

研究生一年级

建筑设计研究（一）：基本设计

· 傅筱

课程类型：必修

学时学分：40 学时／2 学分

Graduate Program 1st Year

DESIGN STUDIO 1: BASIC DESIGN · FU Xiao

Type: Required Course

Study Period and Credits: 40 hours/2 credits

课题内容

宅基地住宅设计

教学目标

课程从“场地、空间、功能、经济性”等建筑的基本问题出发，通过宅基地住宅设计，训练学生对建筑逻辑性的认知，并让学生理解有品质的设计是以基本问题为基础的。

研究主题

设计的逻辑思维

教学内容

在 A、B 两块在宅基地内任选一块进行住宅设计。

Subject Content

Homestead Housing Design

Training Objective

The course starts from fundamental issues of architecture such as “site, space, function, and economical efficiency”, aims to train students to cognize architectural logics, and allow them to understand that quality design is based on such fundamental issues.

Research Subject

Logical thinking of design

Teaching Content

Select one from two homesteads A and B and conduct housing design.

研究生一年级

建筑设计研究（一）：基本设计

· 张雷

课程类型：必修

学时学分：40 学时／2 学分

Graduate Program 1st Year

DESIGN STUDIO 1: BASIC DESIGN · ZHANG Lei

Type: Required Course

Study Period and Credits: 40 hours/2 credits

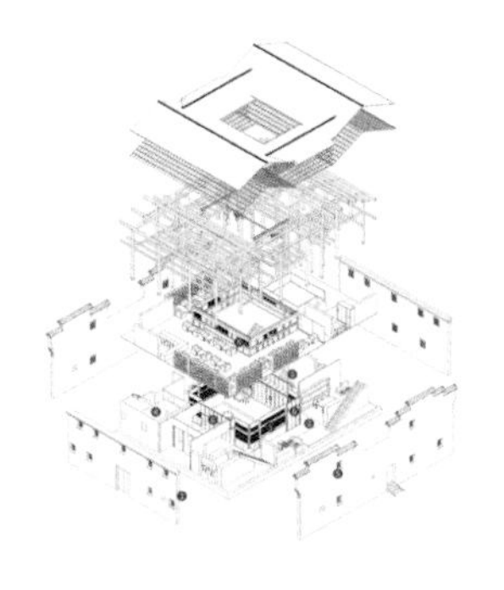

课题内容

传统乡村聚落复兴研究

教学目标

课程从“环境”“空间”“场所”与“建造”等基本的建筑问题出发，对乡村聚落肌理、建筑类型及其生活方式进行分析研究，通过功能置换后的空间再利用，从建筑与基地、空间与活动、材料与实施等关系入手，强化设计问题的分析，强调准确的专业性表达。通过设计训练，达到对地域文化以及建筑设计过程与方法的基本认识与理解。

研究主题

乡土聚落／民居类型／空间再利用／建筑更新／建造逻辑

教学内容

对选定的乡村聚落进行调研，研究功能置换和整修改造的方法和策略，促进乡村传统村落的复兴。

Subject Content

Research on Revitalization of Traditional Rural Settlements

Training Objective

This course starts with basic architectural problems like “environment”, “space”, “site” and “construction”, analyzes and studies the texture, architectural type and life style of rural settlement, strengthens the analysis of design problems and emphasizes accurate professional expression from the relationship between building and base, space and activities, and material and implementation through spatial reuse after function replacement, to obtain basic knowledge and understanding of the regional culture as well as the process and methods of architectural design.

Research Subject

Rural settlement / types of folk house / reutilization of space / building renovation / constructional logic

Teaching Content

Conduct investigation and research on selected rural settlement, study methodology and strategy of function replacement and renovation and improvement, and promote revitalization of traditional rural villages.

研究生一年级

建筑设计研究（一）：概念设计

· 冯路

课程类型：必修

学时学分：40 学时／2 学分

Graduate Program 1st Year

DESIGN STUDIO 1: CONCEPTUAL DESIGN · FENG Lu

Type: Required Course

Study Period and Credits: 40 hours/2 credits

课题内容

半透明性 I

研究主题

以设计为研究，探索“半透明性”的建筑学意义和形式，及其在绘图中的表现方式。半透明性是一种空间关系，它有两种：一种是界面的半透明；一种是结构的，或者折叠的半透明。前者来自于半透明材料，后者建立于空间结构。半透明性的重点在于空间体验，在于遮蔽与揭开，在于对被遮蔽对象的追寻。半透明性因而可以成为一种建筑学的工作方法。

教学内容

以南京花露岗为设计基地和研究对象，基地内包括城墙、愚园、厂房和一大片待开发的荒地。在重新填充这块区域的时候，在设计中如何面对城墙、愚园、厂房以及临时的荒野本身，成为一个值得探索的设计问题。

概念设计将包括城市形态和建筑概念设计两个层面的工作。在整体城市空间格局的基础上选取四个节点展开和深化建筑设计。为了使研究工作更有效，设计将主要针对于城市空间形态结构和建筑空间形式本身。在此基础上，通过设计来理解和思考“半透明性”理论。设计研究成果将通过绘图来表达，因此学习和思考绘图本身也将成为本次课程的重要内容。

Subject Content

Translucency I

Research Subject

This research will carry out research by design, to explore the architectural significance and form of “translucency”, as well as its expression in drawings. Translucency is a spatial relationship, including interfacial translucency and structural or folded translucency. The former comes from translucent materials and the latter is based on spatial structure. Translucency focuses on spatial experience, on covering and uncovering as well as the searching of the covered object. Thus translucency can become a working method in architecture.

Teaching Content

This course takes Nanjing Hualugang Hillock as the design base and research object. When the designer refills this region, it is a design issue worth exploring how to treat the City Wall, Yuyuan Garden, the plants as well as the temporary wasteland. The conceptual design includes two aspects, urban morphology design and architectural conceptual design. The students will choose four nodes based on the overall urban spatial pattern, to perform and deepen the architectural design. To make the research more effective, the design mainly focuses on the structure of urban spatial morphology and the architectural space form. On this basis, the students will understand and think about the theory of “translucency” through design. The design and research result will be expressed in drawings. Therefore, it is also an important part of this course to learn and think about drawing itself.

研究生一年级

建筑设计研究（一）：概念设计

· 鲁安东

课程类型：必修

学时学分：40 学时／2 学分

Graduate Program 1st Year

DESIGN STUDIO 1: CONCEPTUAL DESIGN · LU Andong

Type: Required Course

Study Period and Credits: 40 hours/2 credits

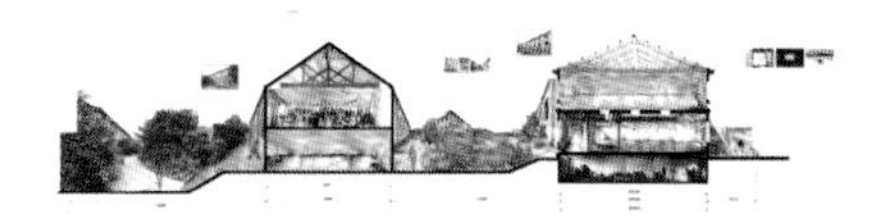

课题内容

扩散：空间营造的流动逻辑

研究主题

以1920—1930年代江浙地区的蚕室建筑为研究对象，关注其对于自然元素的创造性的运用，探索基于环境因素的空间设计方法。

教学内容

本课程从立面和剖面入手，遵循自然的流动逻辑从外向内营造空间，最后完成设计、生成平面。

Subject Content

Diffusion: Flowing Logic of Space Creation

Research Subject

This research takes silkworm nursery of the period 1920s-1930s in Jiangsu and Zhejiang provinces as the research object, pays attention to the creative use of natural elements and explores methods of space design based on environmental elements.

Teaching Content

This course starts from the facade and the profile, creates the space from outside to inside based on natural flow logic and finally completes the design and generates the plan.

研究生一年级

建筑设计研究（二）：建构设计

·傅筱 陈浩如

课程类型：必修

学时学分：40学时／2学分

Graduate Program 1st Year

DESIGN STUDIO 2 : CONSTRUCTIONAL DESIGN · FU Xiao,CHEN Haoru

Type: Required Course

Study Period and Credits: 40 hours/2 credits

课题内容

在浙江莫干山南路乡“60亩农田服务设施规划”场地内实地建造以竹结构为主的“山野乐园”景观小品，供游客和儿童使用。

教学目标

训练学生对设计概念与实地建造的关联性认知。

研究计划

在整个设计研究过程中，将邀请国内知名学者、建筑师前来进行相关学术讲座、研讨和评图；在设计中期进行A、B组评图，选出可建造的方案3~4组，选中与未选中的同学都需将图纸深化到施工图阶段；在最终的实施阶段，所有参与课程的同学都将前往工地参与实地建造，如人手短缺还将增招志愿者。

Subject Content

Field construction of landscape sketch “Garden in the Mountains” featuring bamboo structure on the site of “Service Facility Planning of 60 mu Farmland” in Nanlu Township in Mogan Mountain of Zhejiang, for the tourists and children.

Training Objective

Train the students to perceive the relevance between the design concept and field construction.

Research Plan

During the design and research, we will invite famous scholars and architects in China to give relevant academic lectures, attend the seminars and review the drawings. We will review the drawings in group A and group B in the midterm of design and choose 3~4 buildable proposals. Both the selected and unselected students need to deepen the drawings to the phase of construction drawings. In the final implementation phase, all students attending this class will go to the construction site and participate in the field construction. In the case of shortage of manpower, we will also recruit volunteers.

研究生一年级

建筑设计研究（二）：建构设计

·郭屹民 张准

课程类型：必修

学时学分：40学时／2学分

Graduate Program 1st Year

DESIGN STUDIO 2:CONSTRUCTIONAL DESIGN · GUO Yimin,ZHANG Zhun

Type: Required Course

Study Period and Credits: 40 hours/2 credits

课题内容

南京大学北苑体育中心设计

教学目标

1. 掌握结构设计基础知识，并会进行结构分析和结构设计。

2. 了解结构的材料与建造，并会通过材料和建造进行建筑构造设计。

3. 了解结构设计与功能、场地的关系，并会进行与功能相关的建筑设计。

教学内容

北苑校区现有体育馆设施陈旧、功能单一，已无法满足现有师生更加丰富多彩的健身锻炼需求及社交活动要求。现拟在原体育馆建筑用地范围内改扩建体育中心。

通过这样的课题，学生能够真正理解建筑造型并非是随心所以、为所欲为的个人化行为。在他们所熟知的场地、功能之外，还有着技术层面的约束。这种体会也将使得学生有机会通过结构这一视角反思建筑空间的本质，从而触发他们对建筑更加深入的思考。

Subject Content

Design of the Beiyuan Sports Center of Nanjing University

Training Objective

1. To grasp the basic knowledge of structural design, and to be able to conduct analysis on and design of structure.

2. To understand the materials and construction of structure, and to be able to conduct building structure design with the materials and through construction.

3. To understand relations among structural design, functions and the field, and to be able to conduct function-related architectural design.

Teaching Content

There has an existing gymnasium with outdated facilities and unitary function at Beiyuan campus. It cannot meet the demand of faculty and students on more colorful fitness exercises and social activities. Now it plans to improve and expand the sport center within the land area of the existing gymnasium.

Through this course, students can really understand that architectural modeling is by no means personal behavior that we can do what we want. There are restraints in term of technology in addition to that of field and functions that we are familiar with. Such experience also allows students to have the opportunity to reflect on the nature of architectural space in the view of structure, thus trigger more in-depth thinking on architecture by them.

研究生一年级

建筑设计研究（二）：城市设计

· 丁沃沃

课程类型：必修

学时学分：40 学时／2 学分

Graduate Program 1st Year

DESIGN STUDIO 2: URBAN DESIGN · DING Wowo

Type: Required Course

Study Period and Credits: 40 hours/2 credits

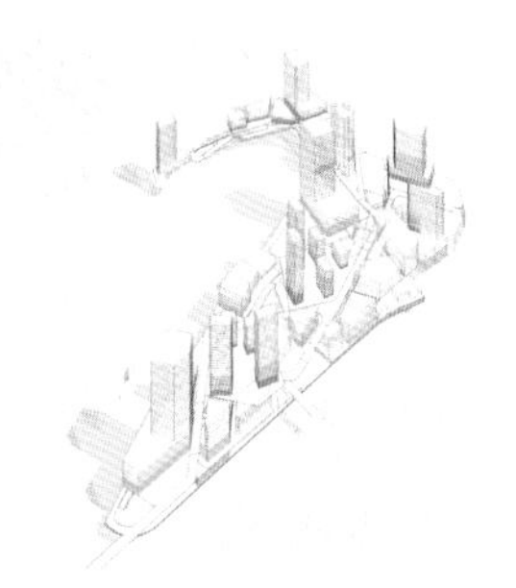

课题内容

城市更新与城市设计

教学目标

我国沿海发达城市城市化进程已经进入一个新的阶段，即由扩张型发展逐渐转为紧凑型发展，城市建设的粗犷型模式即将终结。基于既有建成区的城市更新将成为城市建设的主要途径，改变土地使用性质、提高土地使用效率，以及增大建设密度都将成为建筑师设计工作的主要部分。本课题拟通过基于真实地块的设计实验，认知城市更新中面临的问题，了解城市建筑角色和城市物质空间的本质，初步掌握城市建筑与城市空间塑造之间的关系。此外，通过城市空间设计练习进一步深化空间设计的技能和方法。

关键术语

城市更新、立体交通、空间密度、城市形态

操作元素

“层”“茎”“界”与“空”。

教学内容

1.以南京下关地块作为设计实验的场所，通过场地调研、场地地理条件认知、案例研习、设计分析和设计实验，探讨高效城市空间的建构方法和内涵。

2.通过设计拓展图示技能与表现方法，从空间意向出发建构城市物质空间的层与界面，由“空间意向”“扫描”出城市空间的“层”与“界”。

Subject Content

Urban Renewal and Urban Design

Training Objective

The urbanization of developed coastal cities in China has stepped into a new stage and is gradually transforming from expanding development to compact development, thus extensive urban construction is coming to an end. Urban renewal based on existing built-up area will become a main route of urban construction. The design of architects will mainly focus on change of land use, improvement of efficiency of land use and increase of building density. Based on design experiment of real plot, this subject aims to help the students cognize the problems in urban renewal, understand the nature of the roles of urban architecture and urban material space, and preliminarily master the relationship between unban architecture and urban space formation. In addition, the students will also practice the skills and methods of further deepening the space design through urban space design.

Key Terms

Urban renewal, three-dimensional transportation, spatial density, urban morphology.

Operation Elements

“Level” “stem” “interface” and “space”.

Teaching Content

1. This course takes Xiaguan block of Nanjing as the site for design experiment, to explore construction method and connotation of efficient urban space through site investigation, cognition of site geographical condition, case study, design analysis and design experiment.

2. This course will develop skills and methods of graphic expression through design, construct the level and interface of urban material space from space intention and “scan” the “level” and “interface” of urban space from “space intention”.

研究生一年级

建筑设计研究（二）：城市设计

· 鲁安东

课程类型：必修

学时学分：40 学时／2 学分

Graduate Program 1st Year

DESIGN STUDIO 2 : URBAN DESIGN · LU Andong

Type: Required Course

Study Period and Credits: 40 hours/2 credits

课题内容

南京长江大桥城市设计研究

教学目标

本课程以长江大桥公园改造的实际项目为基础，以国家级江北新区成立为背景，讨论利用“大桥”重新定义城市空间的可能性。

教学内容

本课程将作为长江大桥博物馆的展览内容，并为“长江大桥历史记忆复兴计划”提供基本架构。本课程将以4人一组，从历史、技术、城市、社会四个角度对大桥开展全面和准确的研究，并与多个相关学科的专家和社会媒体进行合作。课程成果计划出版专著一部。

Subject Content

Study on Urban Design of Nanjing Yangtze River Bridge

Training Objective

This course discusses the possibility of redefining urban space with the “Bridge”, based on the actual project of reconstruction of Yangtze River Bridge Garden, in the context of the establishment of national Jiangbei New District.

Teaching Content

The design in this course will be exhibited in Yangtze River Bridge Museum and will provide basic framework for “Plan for Evoking Historical Memory of Yangtze Bridge”. In this class, the students will be divided into several groups, with four students in each group, to carry out comprehensive and accurate research on the Bridge from four perspectives of history, technology, city and society. Besides, the students will cooperate with experts from many related disciplines and the social media. It is planned to publish a monograph with the course achievements.

建筑理论课程
ARCHITECTURAL THEORY COURSES

本科二年级
建筑导论 • 赵辰等
课程类型：必修
学时/学分：36学时/2学分

Undergraduate Program 2nd Year
INTRODUCTION TO ARCHITECTURE • ZHAO Chen, etc.
Type: Required Course
Study Period and Credits:36 hours / 2 credits

课程内容
1. 建筑学的基本定义
第一讲：建筑与设计/赵辰
第二讲：建筑与城市/丁沃沃
第三讲：建筑与生活/张雷
2. 建筑的基本构成
（1）建筑的物质构成
第四讲：建筑的物质环境/赵辰
第五讲：建筑与节能技术/秦孟昊
第六讲：建筑与生态环境/吴蔚
第七讲：建筑与建造技术/冯金龙
（2）建筑的文化构成
第八讲：建筑与人文、艺术、审美/赵辰
第九讲：建筑与环境景观/华晓宁
第十讲：城市肌体/胡友培
第十一讲：建筑与身体经验/鲁安东
（3）建筑师职业与建筑学术
第十二讲：建筑与表现/赵辰
第十三讲：建筑与几何形态/周凌
第十四讲：建筑与数字技术/吉国华
第十五讲：城市与数字技术/童滋雨
第十六讲：建筑师的职业技能与社会责任/傅筱

Course Content
1. Preliminary of architecture
Lect.1: Architecture and design / ZHAO Chen
Lect.2: Architecture and urbanization / DING Wowo
Lect.3: Architecture and life / ZHANG Lei
2. Basic attribute of architecture
(1) Physical attribute
Lect.4: Physical environment of architecture / ZHAO Chen
Lect.5: Architecture and energy saving / QIN Menghao
Lect.6: Architecture and ecological environment / WU Wei
Lect.7: Architecture and construction technology / FENG Jinlong
(2) Cultural attribute
Lect.8: Architecture and civilization, arts, aesthetic / ZHAO Chen
Lect.9: Architecture and landscaping environment / HUA Xiaoning
Lect.10: Urban tissue / HU Youpei
Lect.11: Architecture and body / LU Andong
(3) Architect: profession and academy
Lect.12: Architecture and presentation / ZHAO Chen
Lect.13: Architecture and geometrical form / ZHOU Ling
Lect.14: Architectural and digital technology / JI Guohua
Lect.15: Urban and digital technology / TONG Ziyu
Lect.16: Architect's professional technique and responsibility / FU Xiao

---

本科三年级
建筑设计基础原理 • 周凌
课程类型：必修
学时/学分：36学时/2学分

Undergraduate Program 3rd Year
BASIC THEORY OF ARCHITECTURAL DESIGN
• ZHOU Ling
Type: Required Course
Study Period and Credits:36 hours / 2 credits

教学目标
本课程是建筑学专业本科生的专业基础理论课程。本课程的任务主要是介绍建筑设计中形式与类型的基本原理。形式原理包含历史上各个时期的设计原则，类型原理讨论不同类型建筑的设计原理。
课程内容
1. 形式与类型概述
2. 古典建筑形式语言
3. 现代建筑形式语言
4. 当代建筑形式语言
5. 类型设计
6. 材料与建造
7. 技术与规范
8. 课程总结
课程要求
1. 讲授大纲的重点内容；
2. 通过分析实例启迪学生的思维，加深学生对有关理论及其应用、工程实例等内容的理解；
3. 通过对实例的讨论，引导学生运用所学的专业理论知识，分析、解决实际问题。

Training Objective
This course is a basic theory course for the undergraduate students of architecture. The main purpose of this course is to introduce the basic principles of the form and type in architectural design. Form theory contains design principles in various periods of history; type theory discusses the design principles of different types of building.
Course Content
1. Overview of forms and types
2. Classical architecture form language
3. Modern architecture form language
4. Contemporary architecture form language
5. Type design
6. Materials and construction
7. Technology and specification
8. Course summary
Course Requirement
1. Teach the key elements of the outline;
2. Enlighten students' thinking and enhance students' understanding of the theories, its applications and project examples through analyzing examples;
3. Guide students using the professional knowledge to analysis and solve practical problems through the discussion of examples.

本科三年级
居住建筑设计与居住区规划原理 • 冷天　刘铨
课程类型：必修
学时/学分：36学时/2学分

Undergraduate Program 3rd Year
THEORY OF RESIDENTIAL BUILDING DESIGN AND RESIDENTIAL PLANNING • LENG Tian, LIU Quan
Type: Required Course
Study Period and Credits:36 hours / 2 credits

课程内容
第一讲：课程概述
第二讲：居住建筑的演变
第三讲：套型空间的设计
第四讲：套型空间的组合与单体设计（一）
第五讲：套型空间的组合与单体设计（二）
第六讲：居住建筑的结构、设备与施工
第七讲：专题讲座：住宅的适应性，支撑体住宅
第八讲：城市规划理论概述
第九讲：现代居住区规划的发展历程
第十讲：居住区的空间组织
第十一讲：居住区的道路交通系统规划与设计
第十二讲：居住区的绿地景观系统规划与设计
第十三讲：居住区公共设施规划、竖向设计与管线综合
第十四讲：专题讲座：住宅产品开发
第十五讲：专题讲座：住宅产品设计实践
第十六讲：课程总结，考试答疑

Course Content
Lect. 1: Introduction of the course
Lect. 2: Development of residential building
Lect. 3: Design of dwelling space
Lect. 4: Dwelling space arrangement and residential building design (1)
Lect. 5: Dwelling space arrangement and residential building design (2)
Lect. 6: Structure, detail, facility and construction of residential buildings
Lect. 7: Adapt ability of residential building, supporting house
Lect. 8: Introduction of the theories of urban planning
Lect. 9: History of modern residential planning
Lect. 10: Organization of residential space
Lect. 11: Traffic system planning and design of residential area
Lect. 12: Landscape planning and design of residential area
Lect. 13: Public facilities and infrastructure system
Lect. 14: Real estate development
Lect. 15: The practice of residential planning and housing design
Lect. 16: Summary, question of the test

研究生一年级
现代建筑设计基础理论 • 张雷
课程类型：必修
学时/学分：18学时/1学分

Graduate Program 1st Year
PRELIMINARIES IN MODERN ARCHITECTURAL DESIGN • ZHANG Lei
Type: Required Course
Study Period and Credits:18 hours/1 credit

课程内容
1. 现代设计思想的演变
2. 基本空间的组织
3. 建筑类型的抽象和还原
4. 材料运用与建造问题
5. 场所的形成及其意义
6. 今天的工作原则与策略

建筑可以被抽象到最基本的空间围合状态来面对它所必须解决的基本的适用问题，用最合理、最直接的空间组织和建造方式去解决问题，以普通材料和通用方法去回应复杂的使用要求，是建筑设计所应该关注的基本原则。

Course Content
1. Transition of the modern thoughts of design
2. Arrangement of basic space
3. Abstraction and reversion of architectural types
4. Material application and constructional issues
5. Formation and significance of sites
6. Nowaday working principles and strategies

Architecture can be abstracted to the most fundamental state of space enclosure, so as to confront all the basic applicable problems which must be resolved. The most reasonable and direct mode of space arrangement and construction shall be applied; ordinary materials and universal methods shall be used as the countermeasures to the complicated application requirement. These are the basic principles on which an architecture design institution shall focus.

研究生一年级
现代建筑设计方法论 • 丁沃沃
课程类型：必修
学时/学分：18学时/1学分

Graduate Program 1st Year
METHODOLOGY OF MODERN ARCHITECTURAL DESIGN • DING Wowo
Type: Required Course
Study Period and Credits:18 hours/1 credit

课程内容
以建筑历史为主线，讨论建筑设计方法演变的动因/理念及其方法论。基于对传统中国建筑和西方古典建筑观念异同的分析，探索方法方面的差异。通过分析建筑形式语言的逻辑关系，讨论建筑形式语言的几何学意义。最后，基于城市形态和城市空间的语境探讨了建筑学自治的意义。

1. 引言
2. 西方建筑学的传统
3. 中国:建筑的意义
4. 历史观与现代性
5. 现代建筑与意识的困境
6. 建筑形式语言的探索
7. 反思与回归理性
8. 结语

Course Content
Along the main line of architectural history,this course discussed the evolution of architectural design motivation ideas and methodology.Due to different concepts between the Chinese architecture and Western architecture Matters. The way for analyzing and exploring has to be studied.By analyzing the logic relationship of architectural form language,the geometrical significance of architectural form language is explored. Finally,within the context of urban form and space,the significance of architectural autonomy has been discussed.

1. Introduction
2. Tradition of western architecture
3. Meaning of architecture in China
4. History and modernity
5. Modern architectural ideology and its dilemma
6. Exploration for architectural form language
7. Re-thinking and return to reason
8. Conclusion

研究生一年级
电影建筑学 • 鲁安东
课程类型：选修
学时/学分：36学时/2学分

Graduate Program 1st Year
CINEMATIC ARCHITECTURE • LU Andong
Type: Elective Course
Study Period and Credits:36 hours/2 credits

课程内容
在本课程中，电影被视做一种独特的空间感知和思想交流的媒介。我们将学习如何使用电影媒介来对建筑和城市空间进行微观的分析与研究。本课程将通过循序渐进的教学帮助学生建立一种新的观察和理解建筑的方式，逐步培养学生的空间感知和空间想象的能力，以及使用电影媒介来交流自己的感觉和观点的能力。

本课程将综合课堂授课、实践操作和讲评讨论三种教学形式。在理论教学上，本课程将通过“普遍的运动”、《存在的直觉》《组合的空间》和《城市的幻影》等四讲逐步向学生介绍相关的历史和理论，特别是电影和空间的关系，并帮助学生理解相应的课堂练习。在实践操作教学上，本课程将通过四个小练习和三个大练习，帮助学生从镜头练习到空间表达到观点陈述逐步地掌握电影媒介并将其用于对自己设计能力的培养。而讲评讨论环节将向学生介绍电影媒介的技术和方法，并帮助学生对自己的动手经验进行反思。

Course Content
Film, in this course, is seen as a distinctive medium for the perception of space and the communication of thoughts. We shall learn how to use the unique narrative medium of film to conduct a microscopic study on architecture and urbanism. This course will teach the students of a new way of seeing and knowing architecture. Its purpose is not only to teach theories of urbanism and techniques of filmmaking, but also to teach the students, through a complete case study, of a cinematic (visceral and non-abstract) way of thinking, analyzing and presenting ideas.
This course is composed of a series of 4-hour sessions, which gradually lead the students to undertake their own research project and to produce a cinematic essay on a case study of their own choice. The teaching of this course will be conducted in three forms: the lectures will introduce the students to cinematic ways of seeing and understanding architecture; the tutorials will introduce the students to some basic cinematic techniques, including continuity editing, cinematography, storyboard, shooting script, and post-production; the seminars will review and discuss the students' works in several stages.

城市理论课程
URBAN THEORY COURSES

本科四年级
城市设计及其理论 • 丁沃沃　胡友培
课程类型：必修
学时/学分：36学时/2学分

Undergraduate Program 4th Year
THEORY OF URBAN DESIGN • DING Wowo, HU Youpei
Type: Required Course
Study Period and Credits: 36 hours / 2 credits

课程内容
第一讲：课程概述
第二讲：城市设计技术术语：城市规划相关术语；城市形态相关术语；城市交通相关术语；消防相关术语
第三讲：城市设计方法 —— 文本分析：城市设计上位规划；城市设计相关文献；文献分析方法
第四讲：城市设计方法 —— 数据分析：人口数据分析与配置；交通流量数据分析；功能分配数据分析；视线与高度数据分析；城市空间数据模型的建构
第五讲：城市设计方法 —— 城市肌理分类：城市肌理分类概述；肌理形态与建筑容量；肌理形态与开放空间；肌理形态与交通流量；城市绿地指标体系
第六讲：城市设计方法 —— 城市路网组织：城市道路结构与交通结构概述；城市路网与城市功能；城市路网与城市空间；城市路网与市政设施；城市道路断面设计
第七讲：城市设计方法 —— 城市设计表现：城市设计分析图；城市设计概念表达；城市设计成果解析图；城市设计地块深化设计表达；城市设计空间表达
第八讲：城市设计的历史与理论：城市设计的历史意义；城市设计理论的内涵
第九讲：城市路网形态：路网形态的类型和结构；路网形态与肌理；路网形态的变迁
第十讲：城市空间：城市空间的类型；城市空间结构；城市空间形态；城市空间形态的变迁
第十一讲：城市形态学：英国学派；意大利学派；法国学派；空间句法
第十二讲：城市形态的物理环境：城市形态与物理环境；城市形态与环境研究；城市形态与环境测评；城市形态与环境操作
第十三讲：景观都市主义：景观都市主义的理论、操作和范例
第十四讲：城市自组织现象及其研究：城市自组织现象的魅力与问题；城市自组织系统研究方法；典型自组织现象案例研究
第十五讲：建筑学图式理论与方法：图式理论的研究，建筑学图式的概念；图式理论的应用；作为设计工具的图式；当代城市语境中的建筑学图式理论探索
第十六讲：课程总结

Course Content
Lect.1: Introduction
Lect.2: Technical terms: terms of urban planning, urban morphology, urban traffic and fire protection
Lect.3: Urban design methods — documents analysis: urban planning and policies; relative documents; document analysis techniques and skills
Lect.4: Urban design methods — data analysis: data analysis of demography, traffic flow, public facilities distribution, visual and building height; modelling urban spatial data
Lect.5: Urban design methods — classification of urban fabrics: introduction of urban fabrics; urban fabrics and floor area ratio; urban fabrics and open space; urban fabrics and traffic flow; criteria system of urban green space
Lect.6: Urban design methods — organization of urban road network: introduction; urban road network and urban function; urban road network and urban space; urban road network and civic facilities; design of urban road section
Lect.7: Urban design methods — representation skills of urban Design: mapping and analysis; conceptual diagram; analytical representation of urban design; representation of detail design; spatial representation of urban design
Lect.8: Brief history and theories of urban design: historical meaning of urban design; connotation of urban design theories
Lect.9: Form of urban road network: typology, structure and evolution of road network; road network and urban fabrics
Lect.10: Urban space: typology, structure, morphology and evolution of urban space
Lect.11: Urban morphology: Cozen School; Italian School; French School; Space Syntax Theory
Lect.12: Physical environment of urban forms: urban forms and physical environment; environmental study; environmental evaluation and environmental operations
Lect.13: Landscape urbanism: ideas, theories, operations and examples of landscape urbanism
Lect.14: Researches on the phenomena of the urban self-organization: charms and problems of urban self-organization phenomena; research methodology on urban self-organization phenomena; case studies of urban self-organization phenomena
Lect.15: Theory and method of architectural diagram: theoretical study on diagrams; concepts of architectural diagrams; application of diagram theory; diagrams as design tools; theoretical research of architectural diagrams in contemporary urban context
Lect.16: Summary

研究生一年级
城市形态研究 • 丁沃沃　赵辰　萧红颜
课程类型：必修
学时/学分：36学时/2学分

Graduate Program 1st Year
URBAN MORPHOLOGY RESEACH• DING Wowo, ZHAO Chen, XIAO Hongyan
Type: Required Course
Study Period and Credits: 36 hours / 2 credits

课程要求
1. 要求学生基于对历史性城市形态的认知分析，加深对中西方城市理论与历史的理解。
2. 要求学生基于历史性城市地段的形态分析，提高对中西方城市空间特质及相关理论的认知能力。

课程内容
第一周：序言　概念、方法及成果
第二周：讲座1 城市形态认知的历史基础 —— 营造观念与技术传承
第三周：讲座2 城市形态认知的历史基础 —— 图文并置与意象构建
第四周：讲座3 城市形态认知的理论基础 —— 价值判断与空间生产
第五周：讲座4 城市形态认知的理论基础 —— 钩沉呈现与特征形塑
第六周：讲座5 历史城市的肌理研究
第七周：讲座6 整体与局部 —— 建筑与城市
第八周：讨论
第九周：讲座7 城市化与城市形态
第十周：讲座8 城市乌托邦
第十一周：讲座9 走出乌托邦
第十二周：讲座10 重新认识城市
第十三周：讲座11 城市设计背景
第十四周：讲座12 城市设计实践
第十五周：讲座13 城市设计理论
第十六周：讲评

Course Requirement
1. Deepen the understanding of Chinese and Western urban theories and histories based on the cognition and analysis of historical urban form.
2. Improve the cognitive abilities of the characteristics and theories of Chinese and Western urban space based on the morphological analysis of the historical urban sites.

Course Content
Week 1: Preface — concepts, methods and results
Week 2: Lect. 1 Historical basis of urban form cognition — Developing concepts and transmission
Week 3: Lect. 2 Historical basis of urban form cognition — Apposition of pictures and text and image construction
Week 4: Lect. 3 Theoretical basis of urban form cognition — Value judgement and space production
Week 5: Lect. 4 Theoretical basis of urban form cognition — History representation and feature shaping
Week 6: Lect. 5 Study on the texture of historical cities
Week 7: Lect. 6 Whole and part: Architecture and urban
Week 8: Discussion
Week 9: Lect. 7 Urbanization and urban form
Week 10: Lect. 8 Urban Utopia
Week 11: Lect. 9 Walk out of Utopia
Week 12: Lect. 10 Have a new look of the city
Week 13: Lect. 11 Background of urban design
Week 14: Lect. 12 Practice of urban design
Week 15: Lect. 13 Theory of urban design
Week 16: Discussions

本科四年级
景观规划设计及其理论 • 尹航
课程类型：选修
学时/学分：36学时/2学分

Undergraduate Program 4th Year
LANDSCAPE PALNNING DESIGN AND THEORY
• YIN Hang
Type: Elective Course
Study Period and Credits: 36 hours / 2 credits

课程介绍

景观规划设计的对象包括所有的室外环境，景观与建筑的关系往往是紧密而互相影响的，这种关系在城市中尤为明显。景观规划设计及理论课程希望从景观设计理念、场地设计技术和建筑周边环境塑造等方面开展课程的教学，为建筑学本科生建立更加全面的景观知识体系，并且完善建筑学本科生在建筑场地设计、总平面规划与城市设计等方面的设计能力。

本课程主要从三个方面展开。一是理念与历史：以历史的视角介绍景观学科的发展过程，让学生对景观学科有一个宏观的了解，初步理解景观设计理念的发展；二是场地与文脉：通过阐述景观规划设计与周边自然环境、地理位置、历史文脉和方案可持续性的关系，建立场地与文脉的设计思维；三是景观与建筑：通过设计方法授课、先例分析作业等方式让学生增强建筑的环境意识，了解建筑的场地设计的影响因素、一般步骤与设计方法，并通过与“建筑设计六”和“建筑设计七”的设计任务书相配合的同步课程设计训练来加强学生景观规划设计的能力。

Course Description

The object of landscape planning design includes all outdoor environments; the relationship between landscape and building is often close and interactive, which is especially obvious in a city. This course expects to carry out teaching from perspective of landscape design concept, site design technology, building's peripheral environment creation, etc. to establish a more comprehensive landscape knowledge system for the undergraduate students of architecture, and perfect their design ability in building site design, master plane planning and urban design and so on.

This course includes three aspects:
1. Concept and history;
2. Site and context;
3. Landscape and building.

本科四年级
东西方园林 • 许浩
课程类型：选修
学时/学分：36学时/2学分

Undergraduate Program 4th Year
GARDEN OF EAST AND WEST • XU Hao
Type: Elective Course
Study Period and Credits: 36 hours / 2 credits

课程介绍

帮助学生系统掌握园林、绿地的基本概念、理论和研究方法，尤其了解园林艺术的发展脉络，侧重各个流派如日式园林、江南私家园林、皇家园林、规则式园林、自由式园林、伊斯兰园林的不同特征和关系，使得学生能够从社会背景、环境等方面解读园林的发展特征，并能够开展一定的评价。

Course Description

Help students systematically master the basic concepts, theories and research methods of gardens and greenbelts, especially understand the evolution of gardening, emphasizing the different features and relationships of various genres, such as Japanese gardens, private gardens by the south of Yangtze River, royal gardens, rule-style gardens, free style gardens and Islamic gardens; enable students to interpret the characteristics of garden development from the point of view of social backgrounds, environment, etc. Furthermore to do the evaluation in depth.

研究生一年级
景观规划进展 • 许浩
课程类型：选修
学时/学分：18学时/1学分

Graduate Program 1st Year
LANDSCAPE PLANNING PROGRESS • XU Hao
Type: Elective Course
Study Period and Credits: 18 hours / 1 credit

课程介绍

生态规划是景观规划的核心内容之一。本课程总结了生态系统、生态保护和生态修复的基本概念。大规模生态保护的基本途径是国家公园体系，而生态修复则是通过人为干涉对破损环境的恢复。本课程介绍了国家公园的价值、分类与成就，并通过具体案例论述了欧洲、澳洲景观设计过程中生态修复的做法。

Course Description

Ecological planning is one of the core contents of landscape planning. This course summarizes the basic concepts of ecological systems, ecological protection and ecological restoration. The basic channel of large-scale ecological protection is the national park system, while ecological restoration is to restore the damaged environment by means of human intervention. This course introduces the values, classification and achievements of national parks, and discusses the practices of ecological restoration in the process of landscape design in Europe and Australia.

研究生一年级
景观都市主义理论与方法 • 华晓宁
课程类型：选修
学时/学分：18学时/1学分

Graduate Program 1st Year
THEORY AND METHOD OF LANDSCAPE URBANISM
• HUA Xiaoning
Type: Elective Course
Study Period and Credits: 18 hours / 1 credit

课程介绍

本课程介绍了景观都市主义思想产生的背景、缘起及其主要理论观点，并结合实例，重点分析了其在不同的场址和任务导向下发展起来的多样化的实践策略和操作性工具。通过这些内容的讲授，本课程的最终目的是拓宽学生的视野，引导学生改变既往的思维定式，以新的学科交叉整合的思路，分析和解决当代城市问题。

课程内容

第一讲：导论——当代城市与景观媒介
第二讲：生态过程与景观修复
第三讲：基础设施与景观嫁接
第四讲：嵌入与缝合
第五讲：水平性与都市表面
第六讲：城市图绘与图解
第七讲：AA景观都市主义——原型方法
第八讲：总结与作业

Course Description

The course introduces the backgrounds, the generation and the main theoretical opinions of landscape urbanism. With a series of instances, it particularly analyses the various practical strategies and operational techniques guided by various sites and projects. With all these contents, the aim of the course is to widen the students' field of vision, change their habitual thinking and suggest them to analyze and solve contemporary urban problems using the new ideas of the intersection and integration of different disciplines.

Course Content

Lect. 1: Introduction — contemporary cities and landscape medium
Lect. 2: Ecological process and landscape recovering
Lect. 3: Infrastructure and landscape engrafting
Lect. 4: Embedment and oversewing
Lect. 5: Horizontality and urban surface
Lect. 6: Urban mapping and diagram
Lect. 7: AA Landscape Urbanism — archetypical method
Lect. 8: Conclusion and assignment

历史理论课程
HISTORY THEORY COURSES

本科二年级
中国建筑史（古代）• 刘妍　赵辰
课程类型：必修
学时/学分：36学时/2学分

Undergraduate Program 2nd Year
HISTORY OF CHINESE ARCHITECTURE (ANCIENT)
• LIU Yan, ZHAO Chen
Type: Required Course
Study Period and Credits:36 hours / 2 credits

教学目标

本课程作为本科建筑学专业的历史与理论课程，目标在于培养学生的史学研究素养与对中国建筑及其历史的认识两个层面。在史学理论上，引导学生理解建筑史学这一交叉学科的多种棱面与视角，并从多种相关学科层面对学生进行基本史学研究方法的训练与指导。中国建筑史层面，培养学生对中国传统建筑的营造特征与文化背景建立构架性的认识体系。

课程内容

中国建筑史学七讲与方法论专题。七讲总体走向从微观向宏观，整体以建筑单体-建筑群体-聚落与城市-历史地理为序；从物质性到文化，建造技术-建造制度-建筑的日常性-纪念性-政治与宗教背景-美学追求。方法论专题包括建筑考古学、建筑技术史、人类学、美术史等层面。

Training Objective

As a mandatory historical & theoretical course for undergraduate students, this course aims at two aspects of training: the basic academic capability of historical research and the understanding of Chinses architectural history. It will help students to establish a knowledge frame, that the discipline of History of Architecture as a cross-discipline, is supported and enriched by multiple neighboring disciplines and that the features and development of Chinese Architecture roots deeply in the natural and cultural background.

Course Content

The course composes seven lectures on Chinese Architecture and a series of lectures on methodology. The seven courses follow a route from individual to complex, from physical building to the intangible technique and to the cultural background, from technology to institution, to political and religious background, and finally to aesthetic pursuit. The special topics on methodology include building archaeology, building science and technology, anthropology, art history and so on.

本科二年级
外国建筑史（古代）• 王丹丹　胡恒
课程类型：必修
学时/学分：36学时/2学分

Undergraduate Program 2nd Year
HISTORY OF WESTERN ARCHITECTURE (ANCIENT)
• WANG Dandan,　HU Heng
Type: Required Course
Study Period and Credits: 36 hours / 2 credits

教学目标

本课程力图对西方建筑史的脉络做一个整体勾勒，使学生在掌握重要的建筑史知识点的同时，对西方建筑史在2000多年里的变迁的结构转折（不同风格的演变）有深入的理解。本课程希望学生对建筑史的发展与人类文明发展之间的密切关联有所认识。

课程内容

1. 概论　2. 希腊建筑　3. 罗马建筑　4. 中世纪建筑
5. 文艺复兴建筑（上）6. 文艺复兴建筑（下）
7. 巴洛克建筑　8 近代建筑（上）9. 近代建筑（中）
10. 近代建筑（下）11. 包豪斯　12. 赖特　13. 密斯
14. 康　15. 1960 年代后的建筑　16. 答疑

Training Objective

This course seeks to give an overall outline of Western architectural history, so that the students may have an in-depth understanding of the structural transition (different styles of evolution) of Western architectural history in the past 2000 years. This course hopes that students can understand the close association between the development of architectural history and the development of human civilization.

Course Content

1. Generality　2. Greek Architectures　3. Roman Architectures
4. The Middle Ages Architectures　5. Renaissance Architectures(1)
6. Renaissance Architectures(2)　7. Baroque Architectures
8. 19th-century Architecture(1)　9. 19th-century Architecture(2)
10. 19th-century Architecture(3)　11. Bauhaus　12. Wright
13. Mies　14. Kahn　15. 1970's Architectures　16. Answer Questions

---

本科三年级
外国建筑史（当代）• 胡恒
课程类型：必修
学时/学分：36学时/2学分

Undergraduate Program Program 3rd Year
HISTORY OF WESTERN ARCHITECTURE (MODERN)
• HU Heng
Type: Required Course
Study Period and Credits:36 hours / 2 credits

教学目标

本课程力图用专题的方式对文艺复兴时期的7位代表性的建筑师与5位现当代的重要建筑师作品做一细致的讲解。本课程将重要建筑师的全部作品尽可能在课程中梳理一遍，使学生能够全面掌握重要建筑师的设计思想、理论主旨、与时代的特殊关联、在建筑史中的意义。

课程内容

1. 伯鲁乃涅斯基　2. 阿尔伯蒂　3. 伯拉孟特
4. 米开朗基罗（1）　5. 米开朗琪罗（2）　6. 罗马诺
7. 桑索维诺　8. 帕拉蒂奥（1）　9. 帕拉蒂奥（2）
10. 手法主义　11. 美国现代主义　12. 勒・柯布西耶（1）
13. 勒・柯布西耶（2）　14. 海杜克　15. 日本当代建筑
16. 答疑

Training Objective

This course seeks to make a detailed explanation to the works of 7 representative architects in the Renaissance period and 5 important contemporary architects in a special way. This course will try to reorganize all works of these important architects, so that the students can fully grasp their design ideas, theoretical subject and their particular relevance with the era and significance in the architectural history.

Course Content

1. Brunelleschi　2. Alberti　3. Bramante
4. Michelangelo(1)　5. Michelangelo(2)
6. Romano　7. Sansovino　8. Paratio(1)　9. Paratio(2)
10. Mannerism　11. American Modernism　12. Le Corbusier(1)
13. Le Corbusier(2)　14. Hejduk　15. Contemporary Japanese Architecture　16. Answer Questions

---

本科三年级
中国建筑史（近现代）• 赵辰
课程类型：必修
学时/学分：36学时/2学分

Undergraduate Program 3rd Year
HISTORY OF CHINESE ARCHITECTURE (MODERN)
• ZHAO Chen
Type: Required Course
Study Period and Credits:36 hours / 2 credits

课程介绍

本课程作为本科建筑学专业的历史与理论课程，是中国建筑史教学中的一部分。在中国与西方的古代建筑历史课程先学的基础上，了解中国社会进入近代，以至于现当代的发展进程。

在对比中西方建筑文化的基础之上，建立对中国近现代建筑的整体认识。深刻理解中国传统建筑文化在近代以来与西方建筑文化的冲突与相融之下，逐步演变发展至今天世界建筑文化的一部分之意义。

Course Description

As the history and theory course for undergraduate students of Architecture, this course is part of the teaching of History of Chinese Architecture. Based on the earlier studying of Chinese and Western history of ancient architecture, understand the evolution progress of Chinese society's entry into modern times and even contemporary age.

Based on the comparison of Chinese and Western building culture, establish the overall understanding of China's modern and contemporary buildings. Have further understanding of the significance of China's traditional building culture's gradual evolution into one part of today's world building culture under conflict and blending with Western building culture in modern times.

研究生一年级
建筑理论研究・王骏阳
课程类型：必修
学时/学分：18学时/1学分

Graduate Program 1st Year
STUDY OF ARCHITECTURAL THEORY • WANG Junyang
Type: Required Course
Study Period and Credits:18 hours / 1 credit

课程介绍

本课程是西方建筑史研究生教学的一部分。主要涉及当代西方建筑界具有代表性的思想和理论，其主题包括历史主义、先锋建筑、批判理论、建构文化以及对当代城市的解读等。本课程大量运用图片资料，广泛涉及哲学、历史、艺术等领域，力求在西方文化发展的背景中呈现建筑思想和理论的相对独立性及关联性，理解建筑作为一种人类活动所具有的社会和文化意义，启发学生的理论思维和批判精神。

课程内容

第一讲: 建筑理论概论
第二讲: 建筑自治
第三讲: 柯林・罗：理想别墅的数学与其他
第四讲: 阿道夫・路斯与装饰美学
第五讲: 库哈斯与当代城市的解读
第六讲: 意识的困境：对现代建筑的反思
第七讲: 弗兰普顿的建构文化研究
第八讲: 现象学

Course Description

This course is a part of teaching Western architectural history for graduate students. It mainly deals with the representative thoughts and theories in Western architectural circles, including historicism, vanguard building, critical theory, construction culture and interpretation of contemporary cities and more. Using a lot of pictures involving extensive fields including philosophy, history, art, etc., this course attempts to show the relative independence and relevance of architectural thoughts and theories under the development background of Western culture, understand the social and cultural significance owned by architectures as human activities, and inspire students' theoretical thinking and critical spirit.

Course Content

Lect. 1: Introduction to architectural theories
Lect. 2: Autonomous architecture
Lect. 3: Colin Rowe : the mathematics of the ideal villa and others
Lect. 4: Adolf Loos and adornment aesthetics
Lect. 5: Koolhaas and the interpretation of contemporary cities
Lect. 6: Conscious dilemma: the reflection of modern architecture
Lect. 7: Studies in tectonic culture of Frampton
Lect. 8: Phenomenology

研究生一年级
建筑史研究・胡恒
课程类型：选修
学时/学分：36学时/2学分

Graduate Program 1st Year
ARCHITECTURAL HISTORY RESEARCH • HU Heng
Type: Elective Course
Study Period and Credits:36 hours / 2 credits

教学目标

本课程的目的有二。其一，通过对建筑史研究的方法做一概述，来使学生粗略了解西方建筑史研究方法的总的状况。其二，通过对当代史概念的提出，且用若干具体的案例研究，来向学生展示当代史研究的路数、角度、概念定义、结构布置、主题设定等内容。

课程内容

1. 建筑史方法概述（1）
2. 建筑史方法概述（2）
3. 建筑史方法概述（3）
4. 塔夫里的建筑史研究方法
5. 当代史研究方法 —— 周期
6. 当代史研究方法 —— 杂交
7. 当代史研究方法 —— 阈限
8. 当代史研究方法 —— 对立

Training Objective

This course has two objectives: 1. Give the students a rough understanding of the overall status of the research approaches of the Western architectural history through an overview of them. 2. Show students the approaches, point of view, concept definition, structure layout, theme settings and so on of contemporary history study through proposing the concept of contemporary history and several case studies.

Course Content

1. The overview of the method of architectural history(1)
2. The overview of the method of architectural history(2)
3. The overview of the method of architectural history(3)
4. Tafuri's study method of architectural history
5. The study method of contemporary history — period
6. The study method of contemporary history — hybridization
7. The study method of contemporary history — limen
8. The study method of contemporary history — opposition

研究生一年级
建筑史研究・萧红颜
课程类型：选修
学时/学分：36学时/2学分

Graduate Program 1st Year
ARCHITECTURAL HISTORY RESEARCH • XIAO Hongyan
Type: Elective Course
Study Period and Credits:36 hours / 2 credits

教学目标

本课程尝试从理念与类型两大范畴为切入点，专题讲述中国传统营造基本理念之异变与延续、基本类型之关联与意蕴，强调建筑史应回归艺术史分析框架下阐发相关史证问题及其方法。

课程内容

1. 边角　　2. 堪舆
3. 界域　　4. 传摹
5. 宫台　　6. 池苑
7. 庙墓　　8. 楼亭

Training Objective

This course attempts to start with two areas (concept and type) to state the mutation and continuation, association and implication of basic types of the basic concept of China's traditional construction, emphasizing that the architectural history should return to the art history framework to state relevant history evidence issues and its methods.

Course Content

1. Corner　　2. Geomantic Omen
3. Boundary　　4. Spreading and Copying
5. Table Land of Palace　　6. Pond
7. Temple and Mausoleum　　8. Storied Building Pavilion

建筑技术课程
ARCHITECTURAL TECHNOLOGY COURSES

本科二年级
CAAD理论与实践 • 童滋雨
课程类型：必修
学时/学分：36学时/2学分

Undergraduate Program 2nd Year
THEORY AND PRACTICE OF CAAD • TONG Ziyu
Type: Required Course
Study Period and Credits: 36 hours / 2 credits

课程介绍

在现阶段的CAD教学中，强调了建筑设计在建筑学教学中的主干地位，将计算机技术定位于绘图工具，本课程就是帮助学生可以尽快并且熟练地掌握如何利用计算机工具进行建筑设计的表达。课程中整合了CAD知识、建筑制图知识以及建筑表现知识，将传统CAD教学中教会学生用计算机绘图的模式向教会学生用计算机绘制有形式感的建筑图的模式转变，强调准确性和表现力作为评价CAD学习的两个最重要指标。

本课程的具体学习内容包括：

1. 初步掌握AutoCAD软件和SketchUP软件的使用，能够熟练完成二维制图和三维建模的操作；

2. 掌握建筑制图的相关知识，包括建筑投影的基本概念，平立剖面、轴测、透视和阴影的制图方法和技巧；

3. 图面效果表达的技巧，包括黑白线条图和彩色图纸的表达方法和排版方法。

Course Description

The core position of architectural design is emphasized in the CAD course. The computer technology is defined as drawing instrument. The course helps students learn how to make architectural presentation using computer fast and expertly. The knowledge of CAD, architectural drawing and architectural presentation are integrated into the course. The traditional mode of teaching students to draw in CAD course will be transformed into teaching students to draw architectural drawing with sense of form. The precision and expression will be emphasized as two most important factors to estimate the teaching effect of CAD course.

Contents of the course include:

1. Use AutoCAD and SketchUP to achieve the 2-D drawing and 3-D modeling expertly.

2. Learn relational knowledge of architectural drawing, including basic concepts of architectural projection, drawing methods and skills of plan, elevation, section, axonometry, perspective and shadow.

3. Skills of presentation, including the methods of expression and lay out using mono and colorful drawings

---

本科三年级
建筑技术 1——结构、构造与施工 • 傅筱
课程类型：必修
学时/学分：36学时/2学分

Undergraduate Program 3rd Year
ARCHITECTURAL TECHNOLOGY 1 — STRUCTURE, CONSTRUCTION AND EXECUTION • FU Xiao
Type: Required Course
Study Period and Credits:36 hours / 2 credits

课程介绍

本课程是建筑学专业本科生的专业主干课程。本课程的任务主要是以建筑师的工作性质为基础，讨论一个建筑生成过程中最基本的三大技术支撑（结构、构造、施工）的原理性知识要点，以及它们在建筑实践中的相互关系。

Course Description

The course is a major course for the undergraduate students of architecture. The main purpose of this course is based on the nature of the architect's work, to discuss the principle knowledge points of the basic three technical supports in the process of generating construction (structure, construction, execution), and their mutual relations in the architectural practice.

本科三年级
建筑技术 2——建筑物理 • 吴蔚
课程类型：必修
学时/学分：36学时/2学分

Undergraduate Program 3rd Year
ARCHITECTURAL TECHNOLOGY 2 — BUILDING PHYSICS • WU Wei
Type: Required Course
Study Period and Credits:36 hours / 2 credits

课程介绍

本课程是针对三年级学生所设计，课程介绍了建筑热工学、建筑光学、建筑声学中的基本概念和基本原理，使学生能掌握建筑的热环境、声环境、光环境的基本评估方法，以及相关的国家标准。完成学业后在此方向上能阅读相关书籍，具备在数字技术方法等相关资料的帮助下，完成一定的建筑节能设计的能力。

Course Description

Designed for the Grade-3 students, this course introduces the basic concepts and basic principles in architectural thermal engineering, architectural optics and architectural acoustics, so that the students can master the basic methods for the assessment of building's thermal environment, sound environment and light environment as well as the related national standards. After graduation, the students will be able to read the related books regarding these aspects, and have the ability to complete certain building energy efficiency designs with the help of the related digital techniques and methods.

本科三年级
建筑技术 3——建筑设备 • 吴蔚
课程类型：必修
学时/学分：36学时/2学分

Undergraduate Program 3rd Year
ARCHITECTURAL TECHNOLOGY 3 — BUILDING EQUIPMENT • WU Wei
Type: Required Course
Study Period and Credits:36 hours / 2 credits

课程介绍

本课程是针对南京大学建筑与城市规划学院本科学生三年级所设计。课程介绍了建筑给水排水系统、采暖通风与空气调节系统、电气工程的基本理论、基本知识和基本技能，使学生能熟练地阅读水电、暖通工程图，熟悉水电及消防的设计、施工规范，了解燃气供应、安全用电及建筑防火、防雷的初步知识。

Course Description

This course is an undergraduate class offered in the School of Architecture and Urban Planning, Nanjing University. It introduces the basic principle of the building services systems, the technique of integration amongst the building services and the building. Throughout the course, the fundamental importance to energy, ventilation, air-conditioning and comfort in buildings are highlighted.

研究生一年级
传热学与计算流体力学基础 • 郜志
课程类型：选修
学时/学分：18学时/1学分

Graduate Program 1st Year
FUNDAMENTALS OF HEAT TRANSFER AND COMPUTATIONAL FLUID DYNAMICS • GAO Zhi
Type: Elective Course
Study Period and Credits: 18 hours / 1 credit

课程介绍

本课程的主要任务是使建筑学/建筑技术学专业的学生掌握传热学和计算流体力学的基本概念和基础知识，通过课程教学，使学生熟悉传热学中导热、对流和辐射的经典理论，并了解传热学和计算流体力学的实际应用和最新研究进展，为建筑能源和环境系统的计算和模拟打下坚实的理论基础。教学中尽量简化传热学和计算流体力学经典课程中复杂公式的推导过程，而着重于如何解决建筑能源与建筑环境中涉及流体流动和传热的实际应用问题。

Course Description

This course introduces students majoring in building science and engineering / building technology to the fundamentals of heat transfer and computational fluid dynamics (CFD). Students will study classical theories of conduction, convection and radiation heat transfers, and learn advanced research developments of heat transfer and CFD. The complex mathematics and physics equations are not emphasized. It is desirable that for real-case scenarios students will have the ability to analyze flow and heat transfer phenomena in building energy and environment systems.

研究生一年级
建筑节能与可持续发展・秦孟昊
课程类型：选修
学时/学分：18学时/1学分

Graduate Program 1st Year
ENERGY CONSERVATION AND SUSTAINABLE ARCHITECTURE • QIN Menghao
Type: Elective Course
Study Period and Credits:18 hours / 1 credit

课程介绍

随着我国建筑总量的攀升和居住舒适度的提高，建筑能耗急剧上升，建筑节能成为影响能源安全和提高能效的重要因素之一。建筑节能的关键首先是要设计“本身节能的建筑”，建筑师必须从建筑设计的最初阶段，在建筑的形体、结构、开窗方式、外墙选材等方面融入节能设计的定量分析。而这些很难通过传统建筑设计方法达到，必须依靠建筑技术、建筑设备等多学科互动协作才能完成。这已成为世界各大建筑与城市规划学院教学的一个重点。

本课程将采用双语教学，主要面向建筑设计专业学生讲授建筑物理、建筑技术专业关于建筑节能方面的基本理念、设计方法和模拟软件，并指导学生将这些知识互动运用到节能建筑设计的过程中，在建筑设计专业和建筑技术专业之间建立一个互动的平台，从而达到设计“绿色建筑”的目标，并为以后开展交叉学科研究、培养复合型人才奠定基础。

Course Description

With the rising of China's total number of buildings and the need for living comfort, building energy consumption is rising sharply. Building energy efficiency has become one of the key factors influencing the energy security and energy efficiency. The first key for building energy efficiency is to design “a building that conserves energy itself” and architects must carry out planning at the very beginning of building design. However, it is difficult to satisfy them by means of traditional architectural design approaches; it must be realized by interactive collaboration of diversified subjects including construction technology, construction equipment, etc. Strengthening the interaction of architectural design specialties and construction technology specialties in designing has become a key point in this course as well as in the teaching of various large architecture and urban planning colleges around the world.

研究生一年级
建筑环境学・郜志
课程类型：选修
学时/学分：18学时/1学分

Graduate Program 1st Year
BUILT ENVIRONMENT • GAO Zhi
Type: Elective Course
Study Period and Credits:18 hours / 1 credit

课程介绍

本课程的主要任务是使建筑学/建筑技术学专业的学生掌握建筑环境的基本概念，学习建筑与城市热湿环境、风环境和空气质量的基础知识。通过课程教学，使学生熟悉城市微气候等理论，并了解人体对热湿环境的反应，掌握建筑环境学的实际应用和最新研究进展，为建筑能源和环境系统的测量与模拟打下坚实的基础。

Course Description

This course introduces students majoring in building science and engineering / building technology to the fundamentals of built environment. Students will study classical theories of built / urban thermal and humid environment, wind environment and air quality. Students will also familiarize urban micro environment and human reactions to thermal and humid environment. It is desirable that students will have the ability to measure and simulate building energy and environment systems based upon the knowledge of the latest development of the study of built environment.

研究生一年级
材料与建造・冯金龙
课程类型：必修
学时/学分：18学时/1学分

Graduate Program 1st Year
MATERIAL AND CONSTRUCTION • FENG Jinlong
Type: Required Course
Study Period and Credits:18 hours / 1 credit

课程介绍

本课程将介绍现代建筑技术的发展过程，论述现代建筑技术及其美学观念对建筑设计的重要作用。探讨由材料、结构和构造方式所形成的建筑建造的逻辑方式研究。研究建筑形式产生的物质技术基础，诠释现代建筑的建构理论与研究方法。

Course Description

It introduces the development process of modern architecture technology and discusses the important role played by the modern architecture technology and its aesthetic concepts in the architectural design. It explores the logical methods of construction of the architecture formed by materials, structure and construction. It studies the material and technical basis for the creation of architectural form, and interprets construction theory and research methods for modern architectures.

研究生一年级
计算机辅助技术・吉国华
课程类型：选修
学时/学分：36学时/2学分

Graduate Program 1st Year
TECHNOLOGY OF CAAD • JI Guohua
Type: Elective Course
Study Period and Credits:36 hours / 2 credits

课程介绍

随着计算机辅助建筑设计技术的快速发展，当前数字技术在建筑设计中的角色逐渐从辅助绘图转向了真正的辅助设计，并引发了设计的革命和建筑的形式创新。本课程讲授AutoCAD VBA和RhinoScript编程。让学生在掌握“宏”/“脚本”编程的同时，增强以理性的过程思维方式分析和解决设计问题的能力，为数字建筑设计打下必要的基础。

课程分为三个部分：

1. VB语言基础，包括VB基本语法、结构化程序、数组、过程等编程知识和技巧；

2. AutoCAD VBA，包括AutoCAD VBA的结构、二维图形、人机交互、三维对象等，以及基本的图形学知识；

3. RhinoScript概要，包括基本概念、Nurbs概念、VBScript简介、曲线对象、曲面对象等。

Course Description

Following its fast development, the role of digital technology in architecture is changing from computer-aided drawing to real computer-aided design, leading to a revolution of design and the innovation of architectural form. Teaching the programming with AutoCAD VBA and RhinoScript, the lecture attempts to enhance the students' capability of reasoningly analyzing and solving design problems other than the skills of "macro" or "script" programming, to let them lay the base of digital architectural design.

The course consists of three parts:

1. Introduction to VB, including the basic grammar of VB, structural program, array, process, etc.

2. AutoCAD VBA, including the structure of AutoCAD VBA , 2D graphics, interactive methods, 3D objects, and some basic knowledge of computer graphics.

3. Brief introduction of RhinoScript, including basic concepts, the concept of Nurbs, sammary of VBScript, and Rhino objects.

研究生一年级
GIS基础与应用・童滋雨
课程类型：选修
学时/学分：18学时/1学分

Graduate Program 1st Year
BASIS AND APPLICATION OF GIS • TONG Ziyu
Type: Elective Course
Study Period and Credits:18 hours / 1 credit

课程介绍

本课程的主要目的是让学生理解GIS的相关概念以及GIS对城市研究的意义，并能够利用GIS软件对城市进行分析和研究。

Course Description

This course aims to enable students to understand the related concepts of GIS and the significance of GIS to urban research, and to be able to use GIS software to carry out urban analysis and research.

回声——来自毕业的实践
ECHO—FROM PRACTICES OF GRADUATES

东梓关农居中的“南大建筑”之素养

THE ACCOMPLISHMENT OF “NANJIGN UNIVERSITY ARCHITECTURE”IN THE VILLAGE HOUSES

孟凡浩

土地和成本往往是乡村实践中经常遇到的难题。杭州东梓关乡村农居，作为典型的一次乡村实践，正是如此。每户 120 m^2 的占地面积和单方 1500 元的限价，是设计之初两个不可变的先决条件。

匀质的占地面积限制很容易导致兵营式的“新农居”，如何实现一个如传统村落所具有的丰富多样性的当代村落，是本案面临的最大挑战。在南京大学接受过的“基本设计”教学，让我理解从“环境”“空间”“场所”与“建造”等基本的建筑问题出发，对基本单元、聚落肌理、建筑类型进行分析研究，非常强调单元及单元之间的组合关系，剥离形式本身探究空间原型。在本案的设计中，试图运用这种研究分析方法，从类型学的思考角度抽象共性特点，还原空间原型，尝试通过组织规则以最少的基本单元实现多样性的聚落形态。

设计从基本单元入手，将宅基地轮廓边界与院落边界整合同步考虑，在建筑基底占地面积不超 120 m^2 的前提下，确定了小开间大进深（11 m×21 m）和大开间小进深（16 m×14 m）两种不同方向性的基本单元，建筑基底边界和院落边界形成了一种交织关系，而非传统兵营式布局中宅基地和户内院落的平行关系。两个基本单元建筑基底的适度变化演变出四种类型，将单元通过前后错动、东西镜像形成一个带有公共院落的规模组团，与传统行列式布局相比，在土地节约性、庭院空间的层次性和私密性上都有显著提升。每个规模组团都有一个半公共开放空间，有助于邻里间交往及团体凝聚力和归属感的形成。考虑到村民们对自宅“独立性”的强烈诉求，户与户之间都完全独立，不共用同一堵墙，间距在 1.6~3.2 m 不等。若干个组团的有序生长衍生便逐步发展成有机多样的聚落总图关系，这种从单元生成组团，再由组团演变成村落的生长模式与中国传统古建筑的群体生成关系逻辑一致，也为未来的推广提供了较强的可操作性和可能性。

在基本单元的空间设计中，将使用者生活方式和传统院落情结相结合，注重逻辑的推导分析，通过三个院落串接功能空间，并通过院落界面的不同形成三个透明度完全不一样的院落。前院开敞，内院静谧，后院私密，构建出一个从公共到半公共再到私密的空间序列。

如何在控制造价、降低建造和后期围护难度的同时，在新住宅中延续传统的形式意向是项目的另一难题。本案的设计中，体现了南京大学“建构设计”的一些基本理念和方法：首先，回归到建造的本质，注重建造过程与完成形式之间的逻辑关系，探索工业化模式与传统形式元素之间的关系，选择了砖混结构形式、保温刚性屋面楼板、保温防水外墙以及双层中空玻璃，用白涂料、灰面砖以及仿木纹金属等商品化成熟材料代替木头、夯土、石头等传统材料；在墙体的构造方面，采用 24 厚的砖以不同的砌筑方式形成不同通透度的花格砖墙，对应于楼梯间、设备平台、围墙以及开启扇窗户等处，屋顶檐口设计上以内檐沟做法进行有组织排水，将落水管于“立面”中隐藏；顶部压顶直接由混凝土浇筑出挑，近人尺度的一层挑檐等细节则采用传统的木构工艺建造。通过对传统住宅的形式要素加以提炼与转译，使得所选材料的加工方式得以体现在建造结果中。

本案的设计研究基本是概念、空间、建造这一设计模式在乡建领域的一次物质空间实践尝试，个人感受是，这个项目如同当年南京大学建筑设计训练的一次真题实验。

Land and cost are a big challenge frequently encountered with in countryside practice. The countryside houses in Dongziguan, Hangzhou, as a typical countryside practice, is just like this. The floor space of 120 square meters for each household and the limited price of 1500 yuan for each square meter are the two prerequisites at the beginning of the design.

The even floor space limit is prone to resulting in “new countryside houses” with military camp style. Therefore, how to convert a traditional village into a contemporary village with ample variety is the biggest challenge faced by this case. The “basic design” instruction received from Nanjing University made me start from the basic construction issues like “environment”, “space”, “site”, “construction” and so on, conduct analysis and research on basic unit, human settlement texture and architectural type, put much emphasis on syntagmatic relations of and among units and probe into space prototype by striping away the form itself. In the design of this case, this kind of research and analysis method will be employed to grasp the common characteristics from the perspective of typology, restore space prototype and try to realize diversified human settlement form with the minimum basic units through

organization rules.

The design starts with the basic units, which takes synchronized account of the integration of boundaries of house site and courtyards. On the premise of floor area of building basement not exceeding 120 square meters, two basic units with different directionalities are identified, namely, small width and big length (11m×21m) and big width and small length (16m×14m), and the boundary of building basement interweaves with the boundary of courtyard, which is different from the parallel type of the house site and courtyard in the traditional outline like that of military camp. The moderate transformation of the building basement of the two units gives rise to four types, which forms a sizable cluster with a public courtyard through front and back displacement and east and west symmetry and achieves outstanding promotion in land conservancy and hierarchy and privacy of courtyard space. Each sizable cluster shares a semi-public open space conducive to the formation of cohesiveness and sense of belonging. Taking account of the strong appeal of the villagers for "independence", each house is completely independent without sharing one wall and with 1.6~3.2 meters space between each other. Several groups which grow and develop in good order will gradually develop into organic and diversified general drawing relations of human settlement. The growth pattern which develops from units to groups and from groups to villages is conform to the logic of generating relations of ancient architecture group of traditional China and also provides strong operability and possibility of future promotion.

The space design of the basic units, the lifestyles of the users are combined with traditonal courtyard complex, and logical deduction and analysis is given great attention. Through the join-up function of the three courtyards and the different interfaces of the courtyards, three courtyards with totally different transparency are formed. With the front yard wide-open, the interior yard tranquil and the back yard private, a space sequence ranging from public to semi-public to private is constructed.

When considering how to control cost and reduce the difficulty of construction and later maintenance, there is another big challenge in this project, that is, how to continue with the traditional form in the new house. Some of the basic concepts and methods of the "construction design"of Nanjing University are embodied in this case. First, return to the essence of construction and pay attention to the logic relations between construction process and finished form and explore the relations between industrialized modes and traditional form elements. Choose brick-concrete structure, thermal insulation rigidness roof floor, thermal insulation waterproof outer wall and double glazing and substitute traditonal materials like wood, rammed earth, stone etc. with commercialized mature materials like whiting, gray facing brick and metal with fake wood grain etc.; build walls with bricks of 24 thickness in different masonry styles to form lattice and different transparency so as to be consistent with staircases, facility platforms, enclosure walls and opening windows, etc. . As for the design of cornice on the roof, drain away water with inner eaves gutter and hide downpipe inside the "vetical face"; the coping on the top is achieved with reinforced concrete and the life-size cornice and other details will be constructed with traditonal wood structure techniques. Through the refining and translation of the form elements of traditional houses, the processing methods of those selected materials get access to be reflected in construction results.

The design and research of this case is basically a material space practical trial of the design mode ranging from concept, space and construction in the field of rural construction. My personal feeling is that it is like a test experiment that I did in architecture design training in those years in Nanjing University.

Activity Center
Parking
Parking
Old Village

Parking

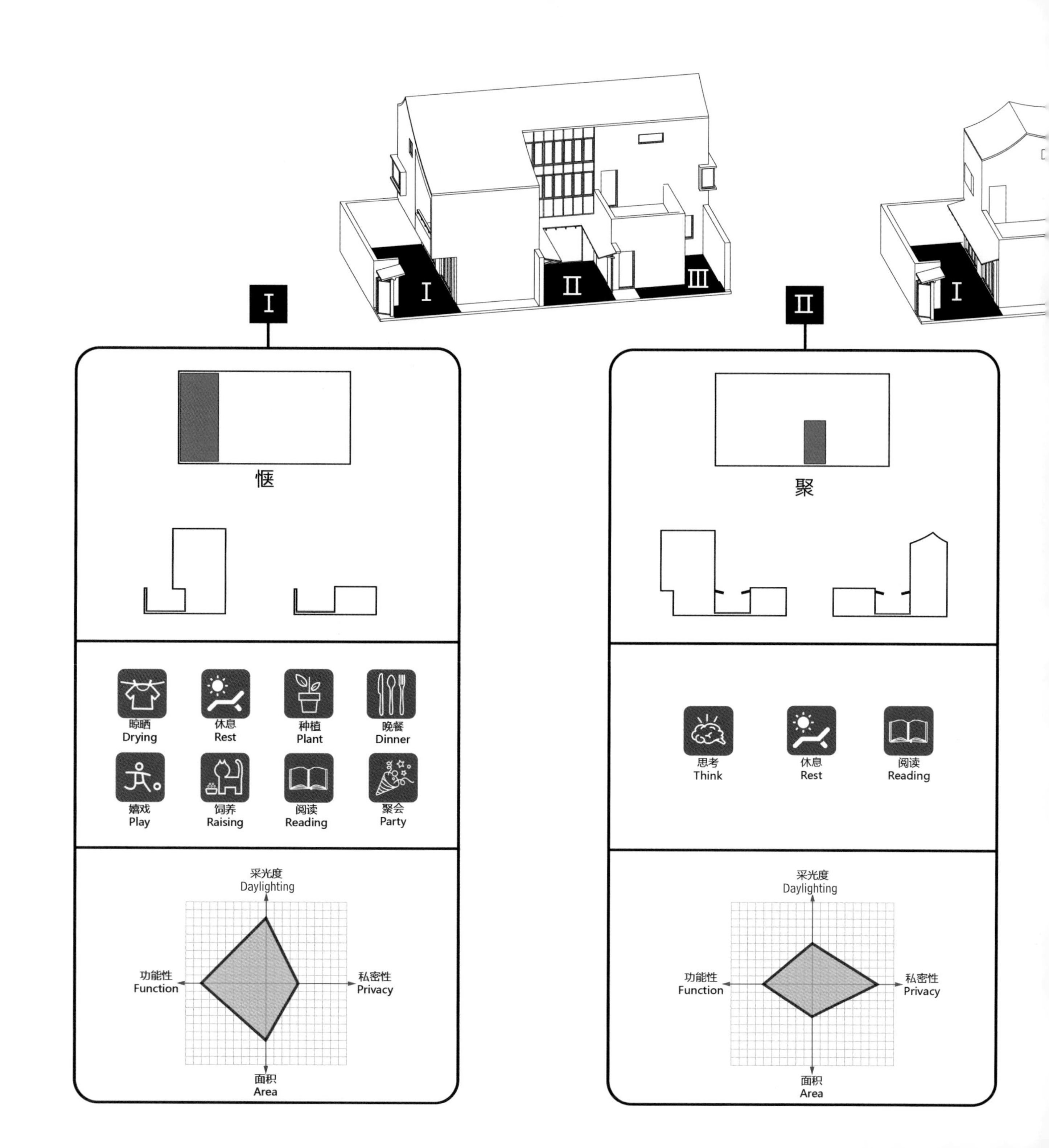

Ⅰ
Ⅱ
Ⅲ
Ⅰ
Ⅰ
惬
晾晒
Drying
休息
Rest
种植
Plant
晚餐
Dinner
嬉戏
Play
饲养
Raising
阅读
Reading
聚会
Party
采光度
Daylighting
功能性
Function
私密性
Privacy
面积
Area
Ⅱ
聚
思考
Think
休息
Rest
阅读
Reading
采光度
Daylighting
功能性
Function
私密性
Privacy
面积
Area

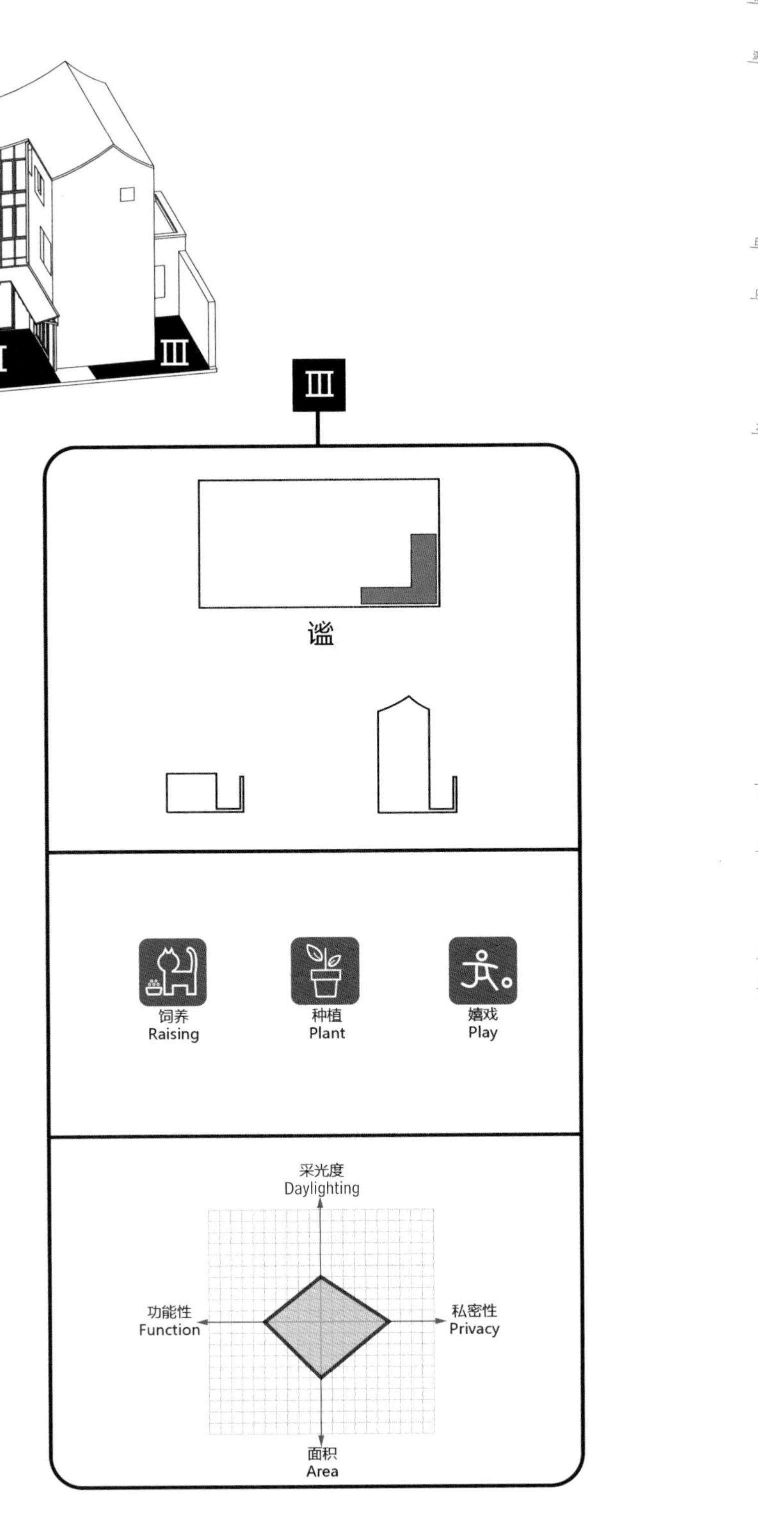

II
III
III
谧
饲养
Raising
种植
Plant
嬉戏
Play
采光度
Daylighting
功能性
Function
私密性
Privacy
面积
Area

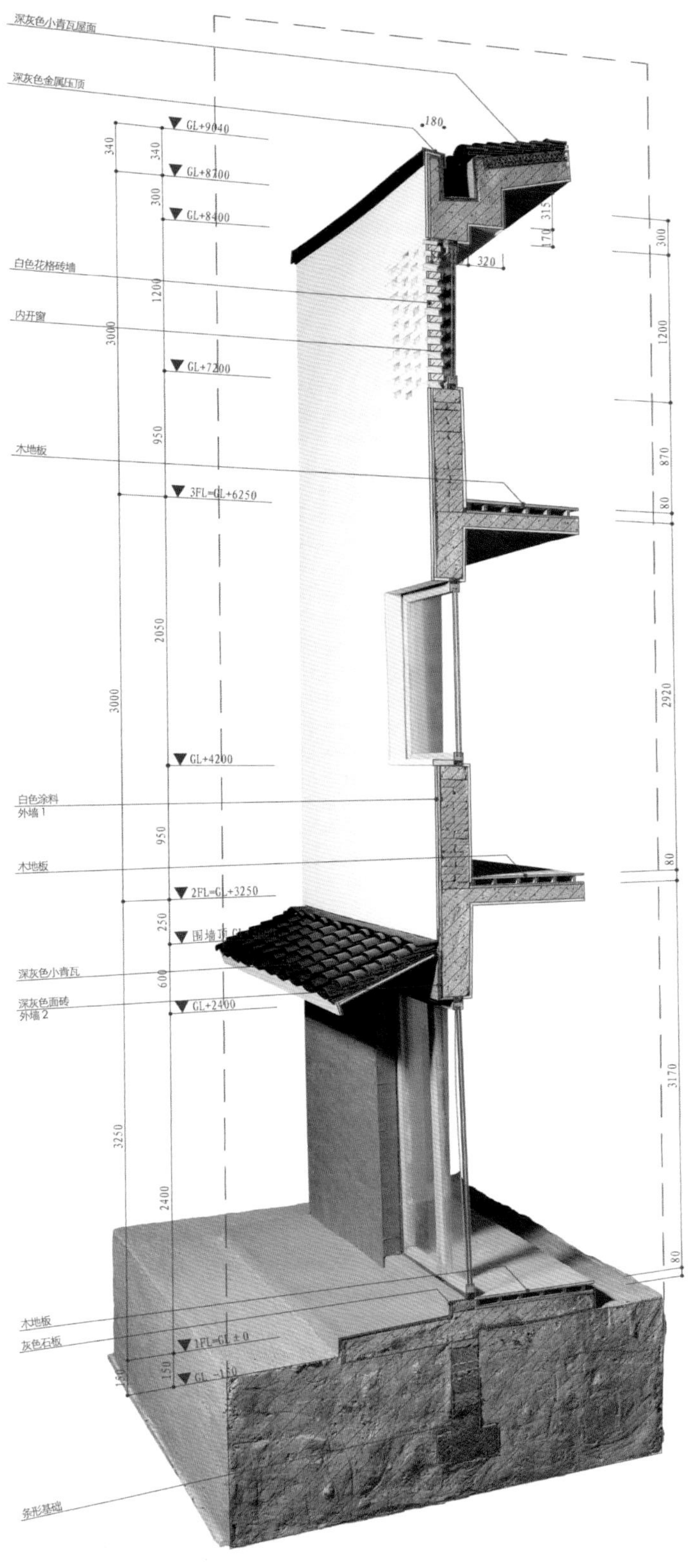

深灰色小青瓦屋面
深灰色金属压顶
180
GL+9040
GL+8700
GL+8400
340
340
300
315
170
300
320
白色花格砖墙
1200
1200
内开窗
3000
GL+7200
950
870
木地板
3FL=GL+6250
80
2050
2920
3000
GL+4200
白色涂料
外墙 1
950
木地板
80
2FL=GL+3250
250
深灰色小青瓦
600
深灰色面砖
外墙 2
GL+2400
3250
2400
3170
80
木地板
灰色石板
1FL=GL±0
150
条形基础

其他
MISCELLANEA

讲座
Lectures

作为研究的实践
——计算思维与数字化工艺视角下的社会生产革新
Practice as a Research: Social Innovation in the view of Computational Thinking and Digital Craftsmanship
袁烽
南京大学蒙民伟楼十楼大教室
二零一六年五月十二日星期四晚七点整

绿色建筑技术讲座
建筑气候热泵技术
Heat Pump Technologies for Building Climate
空气净化与隔离技术
Clean Air & Isolation Technologies
Dr. Ying You
CTO, Infinity Energy Technologies Pty Ltd, Australia
朱枬城
总经理
南京大学蒙民伟楼十楼大教室
二零一六年五月二十日星期五上午九点

《环境的建构》系列讲座二
场地·空间·人——直向建筑近期实践
董功
南京大学蒙民伟楼十楼大教室
二零一五年十一月三日星期二晚七点至十点

CITIARC
山上建筑
营造实践汇报
陈浩如
2015/11/17 星期二 晚上7点
南京大学 蒙民伟楼 十楼1003室

BELLASTOCK
回收建筑材料的再利用
THAT REUSING BUILDING MATERIAL ROCKS!
法国Bellastock实验建造团队全国巡回宣讲
南京大学站
时间 2015年12月13日 周日 19:00
地点 南京大学 蒙民伟楼10楼1003室
主讲人
Antoine Aubinais

建构设计五例
张准
南京大学蒙民伟楼十楼1003室
二零一五年十二月八日星期二晚上七点整

CHANGXIN PENG
AMERICAN ARCHITECTS IN CHINA
1840-1949
ARCHITECTURE, LANDSCAPE, CITY PLANNING
WEDNESDAY
CAMPBELL HALL
近现代中国建筑史课程讲座 7
美国建筑师在中国（1840-1949）
AMERICAN ARCHITECTS IN CHINA, 1840-1949
彭长歆
南京大学鼓楼校区教学楼305
二零一五年十二月八日周二下午19

SHANSI·BUILDING
PROPOSED FOR THE OBERLIN COLLEGE GROUNDS AT OBERLIN OHIO
近现代中国建筑史课程讲座 5
边疆·边缘·边界
——中国近代建筑史研究之现势与走向
刘亦师
南京大学鼓楼校区教学楼305室
二零一五年十一月二十三日周一下午14点

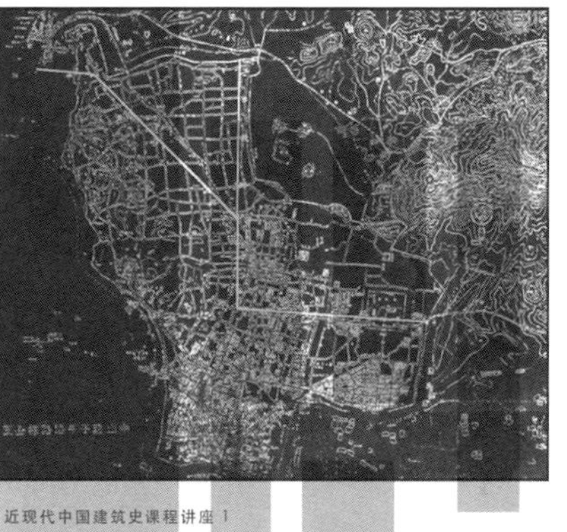

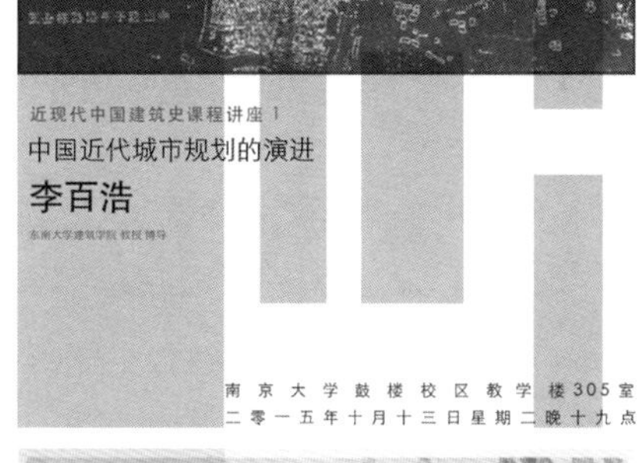
近现代中国建筑史课程讲座 1
中国近代城市规划的演进
李百浩
南京大学鼓楼校区教学楼305室
二零一五年十月十三日星期二晚十九点

近现代中国建筑史课程讲座 6
历史的重“缝”：
1949年后移居香港的中国近代建筑师
MAINLAND ARCHITECTS IN HONG KONG AFTER 1949,
A BIFURCATED HISTORY OF MODERN CHINESE ARCHITECTURE
王浩娱
南京大学鼓楼校区教学楼305室
二零一五年十二月一日周二下午16点

内政部
全国公私建築制式圖案
第四輯
内政部營建司設計繪製
近现代中国建筑史课程讲座 4
规训何以偃旗息鼓？
——建造模式视角下“全国公私建筑制式图案”研究
李海清 博士
南京大学鼓楼校区教学楼305室
二零一五年十一月九日周一下午14点

學習蘇聯先進經驗建設我們的祖國
建设古典主义的乌托邦
——苏联建筑对中国的影响
范思正
南京大学鼓楼校区蒙民伟楼十楼大教室
二零一六年六月一日周三晚七点整

硕士学位论文列表
List of Thesis for Master Degree

研究生姓名	研究生论文标题	导师姓名
陈　逸	无锡市少年宫公共空间设计	张　雷
刘　莹	桐庐莪山畲族乡小住宅设计	张　雷
王　晗	桐庐县莪山畲族乡山哈博物馆设计	张　雷
吴　宾	郑州航空港区幼儿园工程方案设计	张　雷
赵　阳	非规则形体平面布局与建造逻辑研究——以上饶三馆(规划馆)建筑设计为例	张　雷
陈观兴	非透明材料构成孔隙状建筑表皮的设计研究	张　雷
黄龙辉	竹构建筑建造体系研究	张　雷
王淡秋	基于住户需求的住宅产品标准化研究——以万科为例	张　雷
夏　炎	工业化转型下住宅部品体系及其设计方法初探	冯金龙
魏江洋	浅析预制装配式混凝土（PC）技术在民用建筑中的应用与发展	冯金龙
毛军列	传统书院建筑空间在我国现代大学校园建设中的借鉴	冯金龙
力振球	基于人的行为模式心理需求出发的教学楼空间设计——以南京师范大学中北学院理工楼设计为例	冯金龙
蒯冰清	南京师范大学中北学院图书馆建筑方案设计	冯金龙
季　萍	南京理工大学致远楼立面改造设计	冯金龙
戴　波	爱涛天成商务中心方案设计——美术馆展览空间设计研究	冯金龙
徐婉迪	高校学生活动中心功能空间研究——江苏第二师范学院大学生活动中心方案设计	冯金龙
费日晓	建筑立面种植装置设计	吉国华
徐　蕾	基于优化技术的外遮阳形式设计	吉国华
李　彤	基于太阳热辐射的建筑形体生成研究	吉国华
谭子龙	基于建筑风环境分析的Grasshopper与Fluent接口技术研究	吉国华
潘幼建	当代建筑师在中国乡村建造中结构体系的运用研究	傅　筱
奥珅颖	胶合木在当代建筑设计中的应用研究——基于结构角度的设计解析	傅　筱
张　楠	实验楼的空间组织模式研究——以南理工材料实验楼为例	傅　筱
谭发兵	南京汤山集装箱精品酒店设计	傅　筱
姜　智	新防火规范下的社区商业综合体设计	傅　筱
仇高颖	大理下末南村民宿改造设计	周　凌
郭　瑛	元阳地区村落房屋建造体系更新设计	周　凌
贾江南	莫干山清境•原舍二期设计	周　凌
吴超楠	南京门西花露岗地段历史场地研究	周　凌
许　骏	哈尼族传统民居建造体系研究——以云南元阳梯田核心区四个村寨为例	周　凌

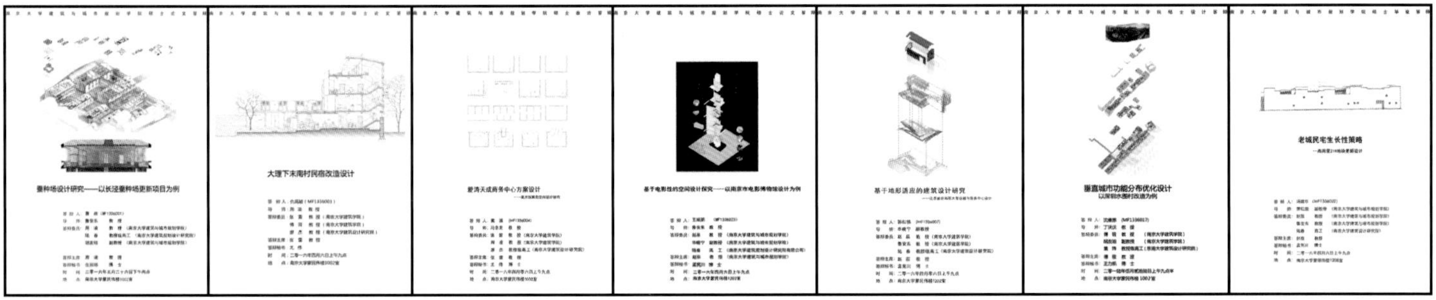

研究生姓名	研究生论文标题	导师姓名
郑金海	乡村建筑改造更新再利用设计研究——以德清县南路村旧村委会建筑为例	周　凌
吴嘉鑫	街廓内地块划分方式与街廓用地指标之间的关系研究——以南京市为例	丁沃沃
沈康惠	垂直城市功能分布优化设计——以深圳水围村改造为例	丁沃沃
郭耘锦	基于地形适应的建筑设计研究——江苏省农科院大型仪器与信息中心设计	华晓宁
施　伟	既有建筑改造中的垂直农业表皮应用及其对室内物理环境的影响研究——江苏省农业科学院农业综合服务中心改造设计	华晓宁
王　倩	江苏省农业科学院主题展览馆设计——建筑水循环与建筑设计的整合研究	华晓宁
徐沁心	旧工业建筑再利用中居住空间植入研究——南京市第二机床厂38号厂房青年公寓改造	华晓宁
孟文儒	从原型到系统的环境营造策略研究	华晓宁
赵倩倩	中国传统建筑檐廊空间及其应用的分析研究——以慈城为例	赵　辰
沙吉敏	场所精神的转换——中国近代城市小住宅与其街廓形式改造修复为纪念性场所的公共空间研究	赵　辰
刘玉婧	闽东北传统建筑现代化更新之室内声舒适度改善研究——以福建政和杨源村水尾民宿设计为例	赵　辰
孙冠成	服务•公共•纪念 ——关于城市街区历史建筑扩建的设计研究以南京利济巷慰安所旧址陈列馆二期建筑设计方案为例	赵　辰
柳筱娴	传统建成环境下交往空间优化——以水斋庵39#—殷高巷24#地块更新设计为例	萧红颜
汤建华	老城民宅生长性策略——南京门西高岗里21#地块更新设计	萧红颜
刘　佳	南京城南门西传统民宅门窗研究	萧红颜
王珊珊	1910s至1930s南通五山片区历史认知分析	萧红颜
许伯晗	南京门西传统院宅空间类型及建造特征浅析	萧红颜
雷冬雪	中国现代新兴建筑类型个案研究——以南京中央医院为例	王骏阳
徐少敏	里下河地区水陆模式转型下的水乡聚落研究	鲁安东
刘彦辰	变迁视野下的中国园林形态分析——以留园为例	鲁安东
符靓璇	环境作用与调控视野下的窗建构体系分析	鲁安东
王斌鹏	基于电影性的空间设计探究——以南京市电影博物馆设计为例	鲁安东
曹　政	蚕种场设计研究——以长泾蚕种场更新项目为例	鲁安东
李　昭	BIM技术辅助的旧办公楼外墙外保温节能改造设计——以正昌办公楼立面改造为例	童滋雨
肖亮平	南京五台山体育馆研究	胡　恒
阮武忠孝	南京市臭氧和PM2.5浓度的变化特征及其室内外（I/O）比例关系研究	郜　志
周荣楼	三维互承结构参数化形式生成初探	童滋雨、吉国华

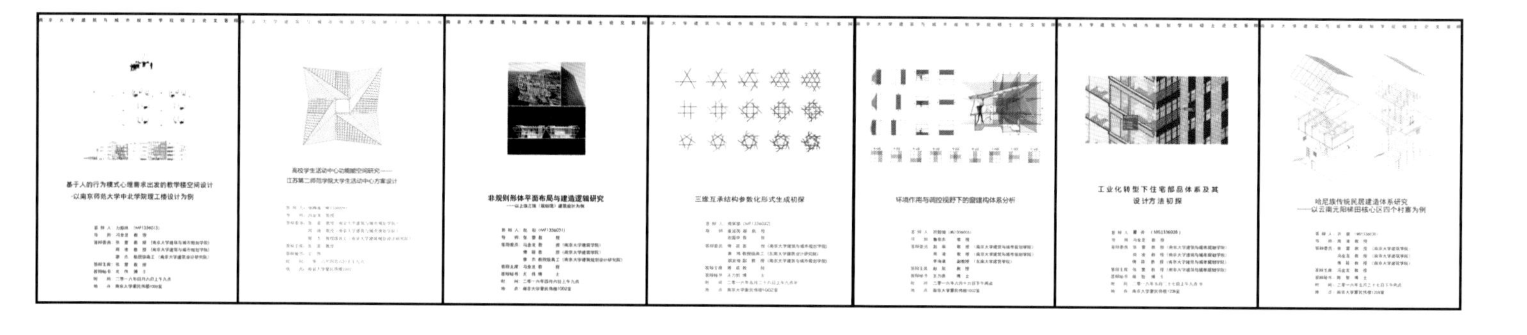

本科生 Undergraduate

2012级学生 Students 2012

陈虹全 CHEN Hongquan
陈思涵 CHEN Sihan
陈 妍 CHEN Yan
从 彬 CONG Bin
段晓昱 DUAN Xiaoyu
高文杰 GAO Wenjie
高祥震 GAO Xiangzhen
葛嘉许 GE Jiaxu
桂 喻 GUI Yu
黄福运 HUANG Fuyun
黄卫健 HUANG Weijian
黄子恩 HUANG Zi'en
季惠敏 JI Huimin
李慧兰 LI Huilan
刘贤斌 LIU Xianbin
刘姿佑 LIU Ziyou
陆怡人 LU Yiren
罗 坤 LUO Kun
钱宇飞 QIAN Yufei
钱雨翀 QIAN Yuchong
全道熏 QUAN Daoxun
沈应浩 SHEN Yinghao
苏 彤 SU Tong
唐林松 TANG Linsong
王 焱 WANG Yan
王一侬 WANG Yinong
吴峥嵘 WU Zhengrong
于明霞 YU Mingxia
臧 倩 ZANG Qian
张馨元 ZHANG Xinyuan
张逸凡 ZHANG Yifan
朱朝龙 ZHU Chaolong
朱凌峥 ZHU Lingzheng
田 甜 TIAN Tian
徐 华 XU Hua
赵媛倩 ZHAO Yuanqian

2013级学生 Students 2013

曹舒琪 CAO Shuqi
陈 露 CHEN Lu
董素宏 DONG Suhong
郭金未 GUO Jinwei
郭 硕 GUO Shuo
贺唯嘉 HE Weijia
胡慧慧 HU Huihui
黄婉莹 HUANG Wanying
黄追日 HUANG Zhuiri
吉雨心 JI Yuxin
贾奕超 JIA Yichao
林之音 LIN Zhiyin
刘稷祺 LIU Jiqi
鲁 晴 LU Qing
罗晓东 LUO Xiaodong
吕 童 Lv Tong
楠田康雄 KUSUDA YASUO
宋宇璕 SONG Yuxun
谭 皓 TAN Hao
涂成祥 TU Chengxiang
王成阳 WANG Chengyang
王 青 WANG Qing
王秋锐 WANG Qiurui
王 瑶 WANG Yao
王智伟 WANG Zhiwei
武 波 WU Bo
夏凡琦 XIA Fanqi
夏 楠 XIA Nan
徐家炜 XU Jiawei
徐瑜灵 XU Yuling
杨 蕾 YANG Lei
章太雷 ZHANG Tailei
赵 焦 ZHAO Jiao
赵梦娣 ZHAO Mengdi
赵中石 ZHAO Zhongshi
周 怡 ZHOU Yi

2014级学生 Students 2014

蔡英杰 CAI Yingjie
曹 焱 CAO Yan
陈妍霓 CHEN Yanni
杜孟泽杉 DU Mengzeshan
胡皓捷 HU Haojie
兰 阳 LAN Yang
梁晓蕊 LIANG Xiaorui
林 宇 LIN Yu
刘 畅 LIU Chang
刘宛莹 LIU Wanying
刘为尚 LIU Weishang
卢 鼎 LU Ding
马西伯 MA Xibo
施少鋆 SHI Shaojun
施孝萱 SHI Xiaoxuan
宋 怡 SONG Yi
宋宇宁 SONG Yuning
宋云龙 SONG Yunlong
唐 萌 TANG Meng
完颜尚文 WANYAN Shangwen
夏心雨 XIA Xinyu
谢 峰 XIE Feng
严紫微 Yan Ziwei
杨云睿 YANG Yunrui
杨 钊 YANG Zhao
尹子晗 YIN Zihan
张 俊 ZHANG Jun
张珊珊 ZHANG Shanshan

2015级学生 Students 2015

卞秋怡 BIAN Qiuyi
吕文倩 LV Wenqian
龚 正 GONG Zheng
顾卓琳 GU Zhuolin
戴添趣 DAI Tianqu
赵 彤 ZHAO Tong
罗紫璇 LUO Zixuan
陈景杨 Chen Jingyang
邸晓宇 DI Xiaoyu
丁展图 DING Zhantu
顾梦婕 GU Mengjie
何 璇 HE Xuan
兰贤元 LAN Xianyuan
李博文 LI Bowen
李心仪 LI Xinyi
刘 博 LIU Bo
刘秀秀 LIU Xiuxiu
刘 越 LIU Yue
罗逍遥 LUO Xiaoyao
毛志敏 MAO Zhimin
秦伟航 QIN Weihang
沈静雯 SHEN Jingwen
汪 榕 WANG Rong
王 晨 WANG Chen
王雪梅 WANG Xuemei
卫 斌 WEI Bin
仙海斌 XIAN Haibin
徐玲丽 XU Lingli
杨鑫毓 YANG Xinyu
杨 洋 YANG Yang
叶庆锋 YE Qingfeng
张昊阳 ZHANG Haoyang
周 杰 ZHOU Jie

研究生 Postgraduate

曹永山 CAO Yongshan
陈成 CHEN Cheng
陈焕彦 CHEN Huanyan
陈娟 CHEN Juan
陈鹏 CHEN Peng
陈中高 CHEN Zhonggao
范丹丹 FAN Dandan
樊璐敏 FAN Lumin
耿健 GENG Jian
韩艺宽 HAN Yikuan
杭晓萌 HANG Xiaomeng
胡绮玭 HU Qipi
胡小敏 HU Xiaomin
黄凯熙 HUANG Kaixi
黄文华 HUANG Wenhua
黄一庭 HUANG Yiting
贾福有 JIA Fuyou
蒋菁菁 JIANG Jingjing
赖友炜 LAI Youwei
李政 LI Zheng
林肖寅 LIN Xiaoyin
林中格 LIN Zhongge
刘赟俊 LIU Yunjun
龙俊荣 LONG Junrong
陆恬 LU Tian
倪绍敏 NI Shaomin
潘东 PAN Dong
邵一丹 SHAO Yidan
司秉卉 SI Binghui
孙燕 SUN Yan
陶敏悦 TAO Minyue
王彬 WANG Bin
王洁琼 WANG Jieqiong
王凯 WANG Kai
王旭静 WANG Xujing
吴黎明 WU Liming
武苗苗 WU Miaomiao
徐怡雯 XU Yiwen
薛晓旸 XUE Xiaoyang
颜晓程 YAN Xiaocheng
杨灿 YANG Can
杨钗芳 YANG Chaifang
杨浩 YANG Hao
杨骏 YANG Jun
杨柯 YANG Ke
殷强 YIN Qiang
余露 YU Lu
俞冰 YU Bing
俞琳 YU Lin
袁亮亮 YUAN Liangliang
岳文博 YUE Wenbo
张成 ZHANG Cheng
张方籍 ZHANG Fangji
张伟 ZHANG Wei
张文婷 ZHANG Wenting
赵芹 ZHAO Qin
赵书艺 ZHAO Shuyi
赵潇欣 ZHAO Xiaoxin
郑国活 ZHENG Guohuo
周青 ZHOU Qing
周雨馨 ZHOU Yuxin
朱鹏飞 ZHU Pengfei
朱煜 ZHU Yu

奥珅颖 AO Shenying
陈观兴 CHEN Guanxing
陈相营 CHEN Xiangying
段艳文 DUAN Yanwen
符靓璇 FU Jingxuan
黄龙辉 Huang Longhui
姜伟杰 JIANG Weijie
雷冬雪 LEI Dongxue
李彤 LI Tong
李招成 LI Zhaocheng
刘佳 LIU Jia
刘彦辰 LIU Yanchen
吕航 Lv Hang
毛军列 Mao Junlie
孟文儒 MENG Wenru
潘柳青 PAN Liuqing
潘幼健 PAN Youjian
沙吉敏 SHA Jimin
谭子龙 TAN Zilong
王淡秋 WANG Danqiu
王冬雪 WANG Dongxue
王珊珊 WANG Shanshan
魏江洋 WEI Jiangyang
吴超楠 WU Chaonan
吴嘉鑫 WU Jiaxin
夏炎 XIA Yan
肖霄 XIAO Xiao
徐少敏 XU Shaomin
许伯晗 XU Bohan
许骏 XU Jun
曹政 CAO Zheng
陈逸 CHEN Yi
仇高颖 QIU Gaoying
戴波 DAI Bo
费日晓 FEI Rixiao
郭瑛 GUO Ying
郭耘锦 GUO Yunjin
季萍 JI Ping
贾江南 JIA Jiangnan
姜智 JIANG Zhi
蒯冰清 KUAI Bingqing
李昭 LI Zhao
力振球 LI Zhenqiu
刘莹 LIU Ying
刘玉婧 LIU Yujing
柳筱娴 LIU Xiaoxian
沈康惠 SHEN Kanghui
施伟 SHI Wei
孙冠成 SUN Guancheng
孙昕 SUN Xin
谭发兵 TAN Fabing
汤建华 TANG Jianhua
王斌鹏 WANG Binpeng
王晗 WANG Han
王倩 WANG Qian
吴宾 WU Bin
徐蕾 XU Lei
徐沁心 XU Qinxin
徐婉迪 XU Wandi
张楠 ZHANG Nan
赵阳 ZHAO Yang
周荣楼 ZHOU Ronglou

车俊颖 CHE Junying
陈博宇 CHEN Boyu
陈凌杰 CHEN Lingjie
陈曦 CHEN Xi
陈晓敏 CHEN Xiaomin
陈修远 CHEN Xiuyuan
程斌 CHENG Bin
高翔 GAO Xiang
顾一蝶 GU Yidie
韩书园 HAN Shuyuan
胡任元 HU Renyuan
黄广伟 HUANG Guangwei
蒋西亚 JIANG Xiya
焦宏斌 JIAO Hongbin
李天骄 LI Tianjiao
廉英豪 LIAN Yinghao
梁万富 LIANG Wanfu
梁耀波 LIANG Yaobo
林陈 LIN Chen
林伟圳 LIN Weizhen
林治 LIN Zhi
刘晨 LIU Chen
刘芮 LIU Rui
刘思彤 LIU Sitong
刘文沛 LIU Wenpei
刘宇 LIU Yu
陆扬帆 LU Yangfan
骆国建 LUO Guojian
宁凯 NING Kai
彭蕊寒 PENG Ruihan
单泓景 SHAN Hongjing
孙雅贤 SUN Yaxian
谭健 TAN Jian
田金华 TIAN Jinhua
王冰卿 WANG Bingqing
王健 WANG Jian
王琳 WANG Lin
王曙光 WANG Shuguang
王政 WANG Zheng
武春洋 WU Chunyang
吴昇奕 WU Shengyi
吴书其 WU Shuqi
吴婷婷 WU Tingting
夏候蓉 XIA HouRong
谢锡淡 XIE Xidan
徐麟 XU Lin
徐思恒 XU Siheng
徐天驹 XU Tianju
许文韬 XU Wentao
徐晏 XU Yan
杨天仪 YANG Tianyi
杨悦 YANG Yue
杨玉菡 YANG Yuhan
姚晨阳 YAO Chenyang
姚梦 YAO Meng
尤逸尘 YOU Yichen
于晓彤 YU Xiaotong
岳海旭 YUE Haixu
查新彧 ZHA Xinyu
张丛 ZHANG Cong
张海宁 ZHANG Haining
张明杰 ZHANG Mingjie
张进 ZHANG Jin
张楠 ZHANG Nan
张强 ZHANG Qiang
郑伟 ZHENG Wei

艾心 AI Xin
曹阳 CAO Yang
陈嘉铮 CHEN Jiazheng
陈立华 CHEN Lihua
陈祺 CHEN Qi
程思远 CHENG Siyuan
种桂梅 CHONG Guimei
迟海韵 CHI Haiyun
崔傲寒 CUI Aohan
方飞 FANG Fei
冯琪 FENG Qi
顾聿笙 GU Yusheng
胡珊 HU Shan
黄凯峰 HUANG Kaifeng
黄丽 HUANG Li
贾福龙 JIA Fulong
江振彦 JIANG Zhenyan
姜澜 JIANG Lan
蒋佳瑶 JIANG Jiayao
蒋建昕 JIANG Jianxin
蒋靖才 JIANG Jingcai
蒋造时 JIANG Zaoshi
黎乐源 LI Leyuan
李若尧 LI Ruoyao
李文凯 LI Wenkai
刘垄鑫 LIU Longxin
刘晓君 LIU Xiaojun
刘泽超 LIU Zechao
柳纬宇 LIU Weiyu
吕秉田 LV Bingtian
缪姣姣 MIAO Jiaojiao
彭丹丹 PENG Dandan
邵思宇 SHAO Siyu
沈佳磊 SHEN Jialei
沈珊珊 SHEN Shanshan
施成 SHI Cheng
宋春亚 SONG Chunya
宋富敏 SONG Fumin
拓展 TUO Zhan
王敏姣 WANG Minjiao
王却奁 WANGQuelian
王晓茜 WANG Xiaoqian
王峥涛 WANG Zhengtao
王子珊 WANG Zishan
吴结松 WU Jiesong
吴松霖 WU Songlin
席弘 XI Hong
谢星宇 XIE Xingyu
谢忠雄 XIE Zhongxiong
徐一品 XU Yipin
徐亦杨 XU Yiyang
杨益晖 YANG Yihui
杨肇伦 YANG Zhaolun
于慧颖 YU Huiying
余星凯 YU Xingkai
张本纪 ZHANG Benji
张豪杰 ZHANG Haojie
张洪光 ZHANG Hongguang
张靖 ZHANG Jing
张黎萌 ZHANG Limeng
张欣 ZHANG Xin
张学 ZHANG Xue
赵婧靓 ZHAO Jingliang
赵伟 ZHAO Wei
周剑晖 ZHOU Jianhui
周明辉 ZHOU Minghui
周贤春 ZHOU Xianchun
周洋 ZHOU Yang
邹晓蕾 ZOU Xiaolei

图书在版编目（CIP）数据

南京大学建筑与城市规划学院建筑系教学年鉴. 2015—2016 / 王丹丹编. -- 南京：东南大学出版社，2016.12
ISBN 978-7-5641-6876-6

Ⅰ. ①南… Ⅱ. ①王… Ⅲ. ①建筑学—教学研究—高等学校—南京市—2015—2016—年鉴②城市规划—教学研究—高等学校—南京—2015—2016—年鉴 Ⅳ. ①TU-42

中国版本图书馆CIP数据核字（2016）第292536号

编 委 会：丁沃沃　赵　辰　吉国华　周　凌　王丹丹
装帧设计：王丹丹　丁沃沃
版面制作：梁晓蕊　刘宛莹　施孝萱　宋　怡
参与制作：颜骁程　陶敏悦
责任编辑：姜　来　魏晓平

出版发行：东南大学出版社
社　　址：南京市四牌楼 2号
出 版 人：江建中
网　　址：http://www.seupress.com
邮　　箱：press@seupress.com
邮　　编：210096
经　　销：全国各地新华书店
印　　刷：南京新世纪联盟印务有限公司
开　　本：787mm×1092mm　1/20
印　　张：9.5
字　　数：595千
版　　次：2016年12月第1版
印　　次：2016年12月第1次印刷
书　　号：ISBN 978-7-5641-6876-6
定　　价：64.00元
